IMAGES
of America

ALASKA'S WHALING COAST

On the Cover: On April 21, 1914, after lying idle for nearly two years, the whaling steamer *Kodiak*, of the newly organized Pacific Sea Products Company, heads for the shore station of the company at Akutan. Capt. Conrad Christiansen commands the *Kodiak*, with Fred Peterson as harpoon gunner, T.F. Ossinger as chief engineer, and Victor Mann as assistant engineer. (Courtesy of the author.)

IMAGES
of America

ALASKA'S WHALING COAST

Dale Vinnedge

ARCADIA
PUBLISHING

ISBN 978-1-4671-3024-0

Published by Arcadia Publishing
Charleston, South Carolina

Printed in the United States of America

Library of Congress Control Number: 2013933754

For all general information, please contact Arcadia Publishing:
Telephone 843-853-2070
Fax 843-853-0044
E-mail sales@arcadiapublishing.com
For customer service and orders:
Toll-Free 1-888-313-2665

Visit us on the Internet at www.arcadiapublishing.com

This book is dedicated to a friend I met during the summer of 1985 when I visited the Kendall Whaling Museum in Sharon, Massachusetts. I met a visiting scholar from Germany, Klaus Barthelmess, who was performing a summer internship. Over the years, Klaus has never failed to support me and my interests. In whatever city we were in, we went "whaling"—meaning a search for books, magazines, artifacts, or whatever could be acquired. I could tell Klaus where I was traveling, and he would tell me off the top of his head where to go and who to see. In the world of whaling, Klaus was a scholar par excellence and a master of languages. Greasy luck, Klaus.

Contents

ACKNOWLEDGMENTS

The subject of whaling has been a longtime pursuit for me. Through the years, many people have influenced my research and collecting of whaling information and images. Fortunately, I met a few key individuals who shared their passions with me and others who put up with my storytelling.

Dieter Kusgen introduced me to the world of collecting images as history. Whether it be postcards or photographs, it takes special talent to find them where they are and, as time changes, their locations. Without Kusgen's help, I could not have amassed a collection of a size needed to illustrate a book of this nature.

I wish to acknowledge David Hull, past principal librarian of the San Francisco Maritime Museum Library. He would never fail to provide some information or perspective I was not thinking about; his knowledge of the library and maritime contacts were of great benefit. I would like to thank the Nores for opening and sharing their Alaska photographic collection. I also wish to acknowledge the following people for their assistance with their archives: Anjuli Grantham, curator of collections, and Alice Ryser, archivist with the Baranov Museum; Patuk Glenn, Inupiat Heritage Center manager, and Muriel Hopson, oral historian, of the Inupiat Heritage Museum in Barrow, Alaska, for sharing the photographs of the Inupiat History Language and Culture Division; and Keri Koehler, collections manager at San Francisco Maritime National Historical Park.

I would like to thank the Steve Oomittuk, mayor of Point Hope, for an enlightening tour of his town. I would like to extend a special thank you to Marlene Beam, assistant principal, and the faculty and high school students of the Tikigaq High School in Point Hope, Alaska, for sharing their images and their stories of whaling and allowing me to share mine with them.

All images, unless otherwise noted, are courtesy of the author.

Introduction

Whaling in Alaska was being prosecuted thousands of years ago, and this whaling has continued through to this year. The last chapter of whaling in Alaska is currently being played out between the whales and today's Inuit and Inupiat native whalers.

For millennia, bowhead and beluga whales have been migrating north from the Pacific Ocean to the Arctic Ocean during the spring. The whales navigate north as leads (passageways) develop in ice that formed in the sea during the winter, returning in open water (no ice) in the fall. These whales have been hunted for thousands of years by indigenous peoples. Along the coastlines of Alaska, Inuit and Inupiat natives in skin boats (umiaks) have attacked the whales with spears and lances for at least 2,000 years. Their methods are steeped in custom and tradition and have changed little over the centuries until recently.

After 1850, their whaling methods—and, in fact, their entire livelihood—would be changed forever. The bark *Superior* found a large number of whales in the Arctic Ocean. In just a few years, many whalers and almost the entire Pacific fleet came north around June of each year to find a path through the melting ice. The whalers spread out over the Arctic Ocean and began to have contact with the natives who lived along the coast of Alaska. Whalers from America and Europe brought goods to trade (beads, cloth, and so on) but also brought guns, ammunition, and alcohol. The whalers unknowingly also brought illnesses such as measles and venereal diseases, which the natives had no natural defense for. The trade in guns and alcohol was so pervasive that the US government commissioned the revenue cutter *Bear* to protect the whalers in Alaskan waters from other countries and to police the illegal traffic in firearms and alcohol. The whalers built trading posts, some of which, such as Point Barrow, became small towns as the natives took up residence nearby.

The bowhead whales yielded baleen used in the manufacture of corset stays, buggy whips, and more, and in 1905, it was worth nearly $5 per pound. But around the same time, petroleum products began to replace baleen with early plastics that were far cheaper and less dangerous to acquire. The market for baleen practically disappeared, leaving only a few ships in the whaling and trading business. A few ships managed to whale into the 1920s.

A new type of whaling was introduced in Norway in the 1860s. Svend Foyn, a Norwegian, invented the harpoon cannon and the steel whale catcher ship, which enabled the whaling industry to go after species of whales previously ignored as unattainable. This new type of whaling was centered around meat and oil, and the industry was soon resurrected. The Norwegians began to exploit whales in all oceans of the world, and around 1910, whaling companies began to show up in southern Alaska. They built onshore whaling stations and hunted the whales into the 1930s and 1940s; this industry stopped before World War II and did not start up again after the war.

The Inuit and Inupiat whalers continue to go whaling from approximately 15 small towns scattered along the Arctic Ocean and some of the small islands in the Bering Strait. For these natives, whaling is a life-or-death proposition in a land considered uninhabitable by many; without the whale, whole villages probably could not continue to live as basically as they have for many years.

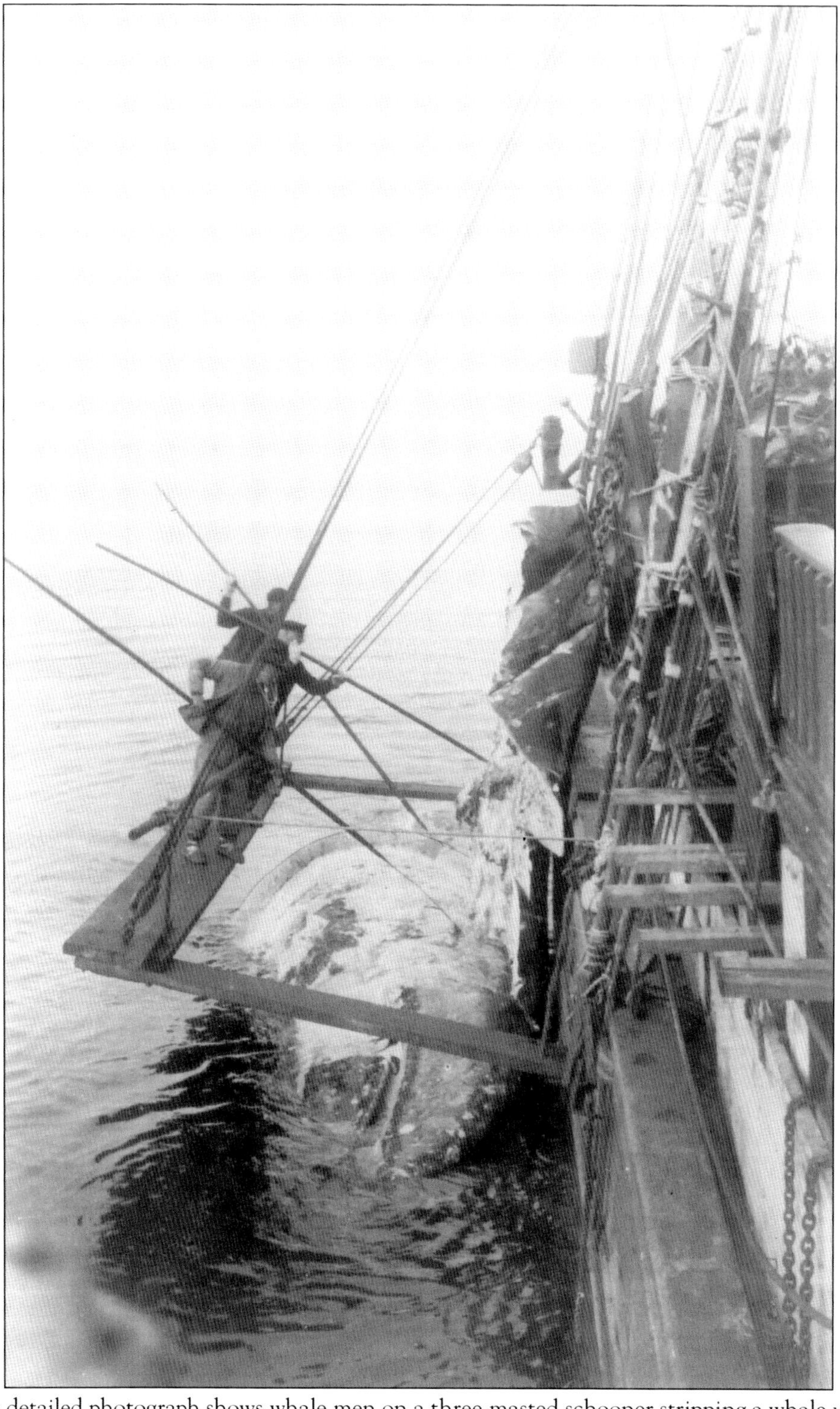

This detailed photograph shows whale men on a three-masted schooner stripping a whale upon the starboard side of the ship.

One

Early Whaling in Alaska

The earliest known whaling in Alaska was from Point Hope in the northwest corner of the state and goes back, judging by carbon dating, about 2,000 years. The village of Point Hope has been continuously occupied during this length of time. It is hard to accurately pinpoint how long other communities have been whaling, because the native Inupiat people are nomadic, with some towns having been recently established by natives who moved there from different places.

Whaling methods remained basically unchanged from early times until locals made contact with commercial whalers who arrived in the mid-19th century and showed the natives more efficient equipment and processes. With only a few changes, Point Hope methods have managed to remain the least changed among the villages that are still whaling today.

There are currently two whaling seasons in Alaska. The first—the spring season—is practiced along the edges of the frozen ice as leads develop in the ice. As the winter ice between Point Hope and the eastern Russian territory begins to break up and melt, the area slowly becomes navigable for the bowhead and beluga whales. These whales navigate from the northern Pacific Ocean through the straits into the Arctic Sea, passing Point Hope at the northwestern point of Alaska.

The second whaling season—the fall season—occurs when the whales that have spent the summer months in the Arctic Ocean, to the northeast of Alaska, begin to return along the northern Alaska coast to warmer waters in the Pacific Ocean. Whaling in the fall season is very different from the spring season, because the ice has melted and therefore the whaling methods need to be different. Point Barrow and a few other small villages in eastern Alaska are the only villages that pursue the whales in the fall. Point Barrow is the only village in which whalers pursue the whales in both the spring and fall seasons.

Prior to each whaling season, preparation includes traditional ceremonies and practices. A whale's tail preserved from the previous season is distributed in a ceremonial fashion with all families of the community receiving portions of it. The members of the whaling crew begin to cut up the tail; the captain receives a strip about four inches wide, then each crew member gets long strips from the side of the tail. The remaining skin and blubber are cut up, and while the cutting is occurring, the captain relates stories of past whaling seasons, much to the enjoyment of those present. After the stories are told, the captain and crews sit down and are served their meals first. Everyone then passes their plates forward for their share of meat and blubber (muktuk). Finally, people form a line to get a share of the whale tail to take home.

Later, near the start of the season, the whaling captain and his crew begin to clean and repair the whaling equipment. This basic equipment consists of the skin boat (umiak), paddles, a pair of oars, two or three sealskin floats, heavy rope, harpoons, and more. The paddles and oars are scraped clean to make them look bright and new. This was done as a throwback to aboriginal times, when it was thought that clean whaling equipment would please the whale and it would allow itself to be killed. The sealskin floats are inflated and any other equipment that is used is checked and made ready. In aboriginal days, the Inuit used harpoons with large ivory toggle heads to strike the whales, and they killed the whales with long, flint-headed lances.

During the early spring, each crew must attach the skins to its umiak to prepare for the season. Crews sometimes choose different seasons to prepare the umiaks, allowing enough time for them to dry and bleach in the sun.

In the past, a boat captain held an important position in the community, as he was normally the wealthiest man in the large extended family that characterized village social structure. The captain was normally a shaman and drew his crew from his family, in which he exercised some authority. The captain selected the place along the ice lead where the boat would be launched. When selecting a campsite, he was careful to choose one on solid ice and not near the edge of the ice, which may break and carry away boats and equipment.

The whalers set up camp, which, in aboriginal times, did not include tents, sleeping bags, or shelter of any kind. They had a small container of water and shared a container of meat that was sent out from the shore. The whalers were not allowed a change of clothing and remained on the ice until the end of whaling season. A south wind would close the leads, which signaled the end of whaling. The equipment was then moved to a solid ice location near the lead and locked up to keep the dogs from eating the sealskin floats. This was done in case the lead might open up again, and whaling could resume quickly.

When a whale was sighted far out in the sea, all the crews launched their boats and started pursuit. If a whale surfaced close to one crew, that crew alone would pursue it. If the first crew missed in its attempt to strike the whale, other crews could join the pursuit. As described in VanStone's *Point Hope, an Eskimo Village in Transition*, the harpooner, sitting in front of the boat, "would stand and drive the harpoon deep into the animal's body." The whale then would sound deep into the water, taking with it the harpoon, attached line, and sealskin floats. The other whale boats would come to the vicinity of where the whale sounded and wait for the floats to come up, which indicated the surfacing of the whale. When the whale surfaced, the other whalers rushed in and attempted to attach more harpoons and lines with floats. Eventually, if killed, the whale would appear on the surface; if merely wounded, it might escape and disappear under the ice.

The crews would tow the dead whales to the nearest point on the ice and cut the tail flukes off for towing—these would be saved for the spring feast. All of the crews participated in hauling the whale onto the ice after a rope had been tied to the tail. The whale was then towed to a firm spot on solid ice where it could be butchered. A runner went to town to let the villagers know a whale had been brought in, and the whole village would come to watch and participate in the butchering process.

The whaler's share of the choice meat, blubber, and whale skin was determined based on each whaler's individual participation in the hunt. The first to strike the whale got the choicest parts. Crews not in on the catch still received a share based upon the timing of their arrival on the scene of the dead whale and touching the whale with their paddle. This distribution method was very elaborate and provided whale products for every household in the community.

At the end of the season, when much of the whale had been distributed, the town would turn out for a celebration of the successful hunt. The captains and their crews were honored in traditional ceremonies—the villagers would play games and take part in the traditional blanket toss, and a grand time was had by all.

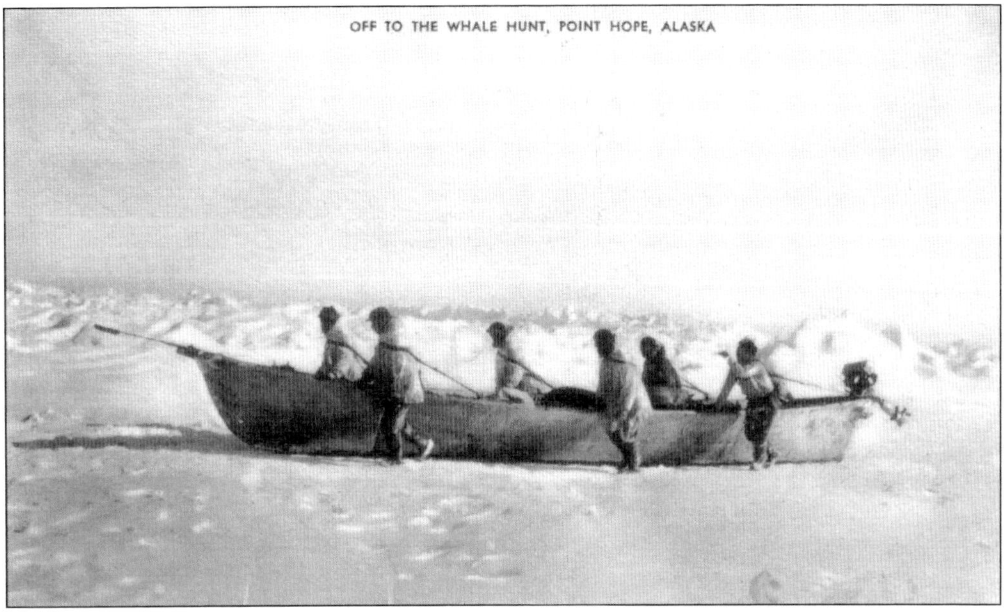

Point Hope whalers walk their umiak across the ice to set up camp for whaling. Point Hope does not participate in the fall whaling, because there is no ice and the whales travel farther offshore. Today, snowmobiles and four-wheel-drive vehicles are used to move the boats and to travel back and forth for supplies.

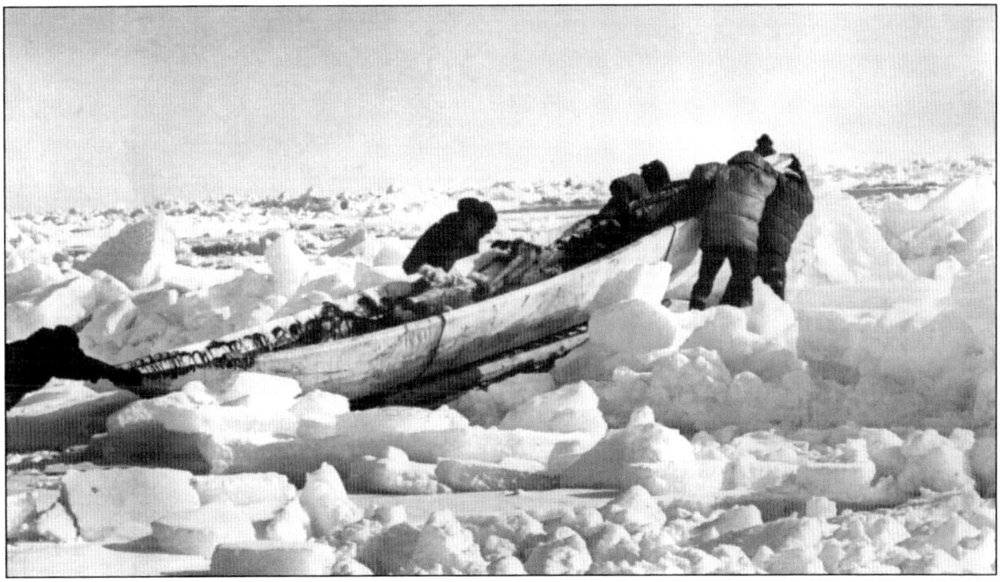

These whalers are pushing an umiak through the large ice boulders toward open water to begin the hunt. The whalers live and hunt on the ice during the spring season. Although they may be three to four miles out to sea on the ice, the whalers hide the umiak behind ice boulders so the whales cannot see them. The bowhead whale can be very wary and may not come near if spooked. Once the captain chooses the spot to launch the boat, it is drawn to the edge of the ice with the bow projecting out over the water. All the equipment is made ready so that when a whale is sighted, there will be no delay in getting into the water. The whaling crews scatter across two to three miles along the ice approximately two to three miles from land.

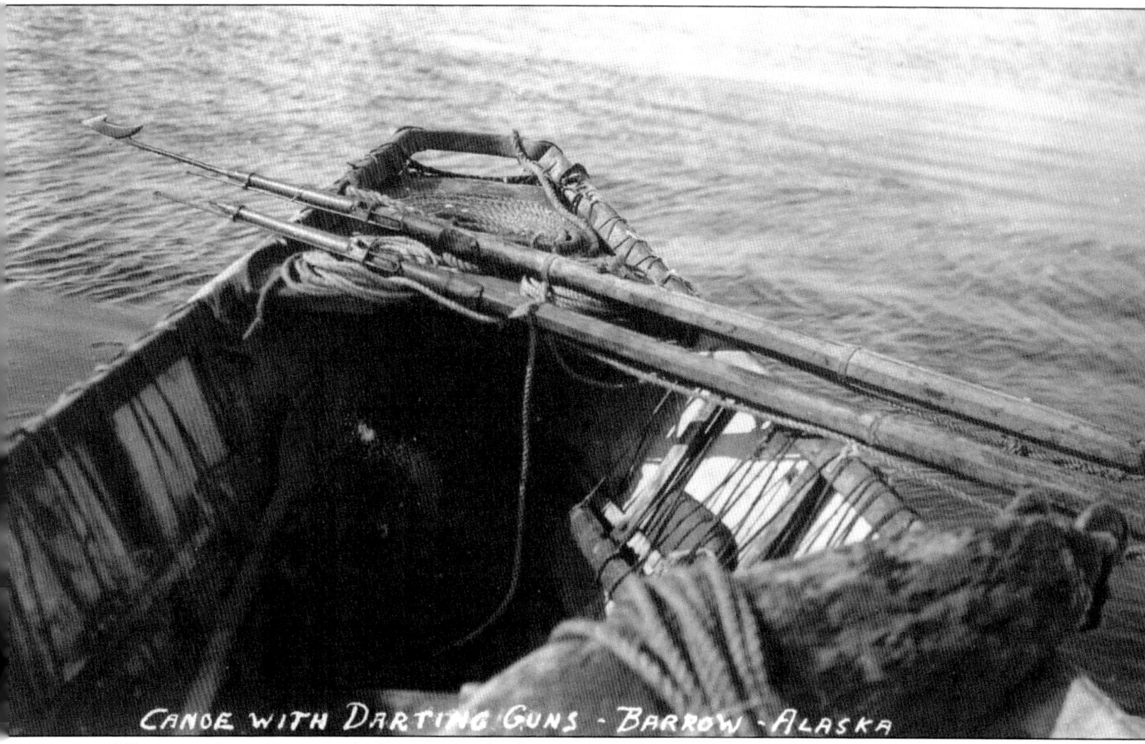

The umiak is a boat with skins stretched over an internal wood frame. They are normally remade on the same frame at the start of every season. The whole process takes about five hours. While the women sew, the men assemble the boat frame. The men replace any broken wood on the boat frame, replace nails, and repair any damage before the skins are put on. The boats are then stored on a stilt frame rack, often made from whale rib bones, at the captain's house. Finally, all of the skins are coated with blubber to help provide some waterproofing. This picture of an umiak includes the whaler's main tools of the trade—the harpoon is thrown first to secure the boat to the whale, then the darting gun is used. The darting gun has an explosive head that enters the whale, and the charge is set off after a few seconds, ideally killing the whale with one dart.

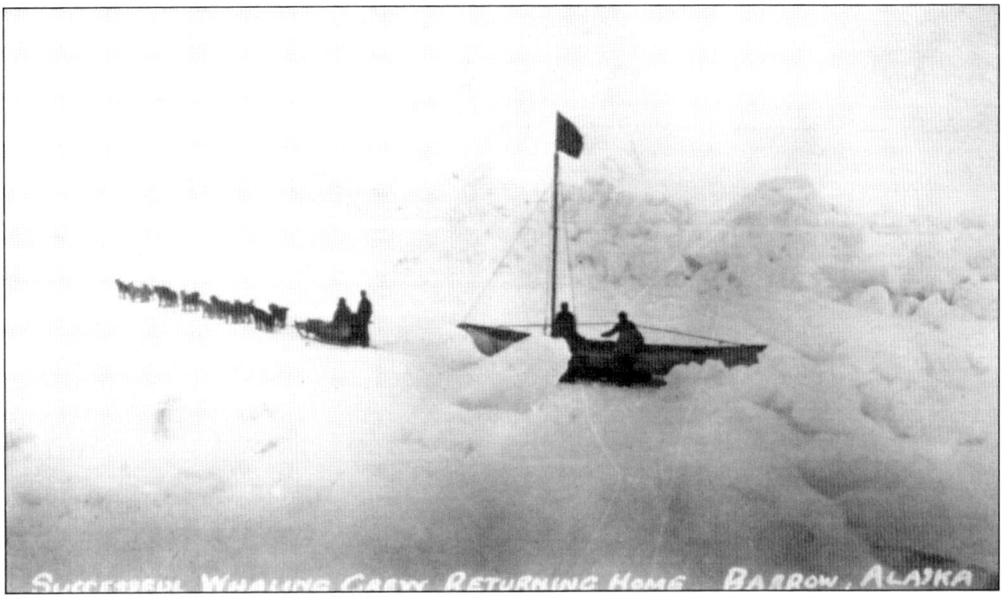

This image shows a crew that has successfully caught a bowhead whale and is towing it to a designated place for flensing. After the crews have towed the whale to the edge of the ice and once the whale has been lanced, the first boat stays back to allow the other boats to come in. The first boat also gets fast to the whale, then all the crews help tow the whale in, with the lead boat containing the crew that originally struck the whale. The successful crews raise a flag and head toward a time of celebration.

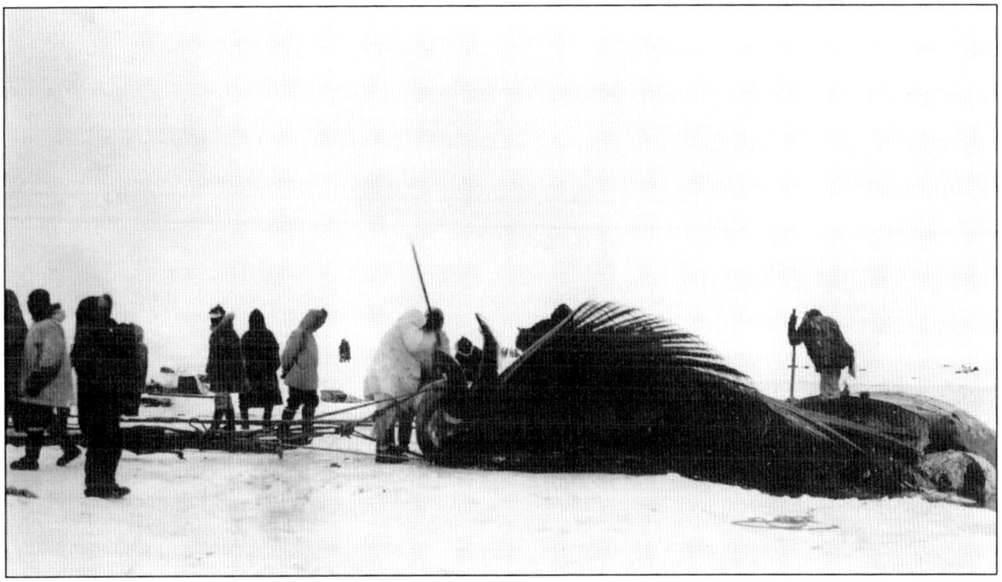

These whalers are preparing to remove the baleen from a whale. Baleen from the bowhead whale was sought after by whalers after the bowheads were discovered in the Arctic in 1848 by the bark *Superior.* The *Superior* was the first ship to maneuver through the melting ice. The price of whalebone reached a peak of just over $5 per pound in 1905 before beginning to fall off until in a short few years it was no longer economically feasible for taking, although it was taken occasionally as a secondary product.

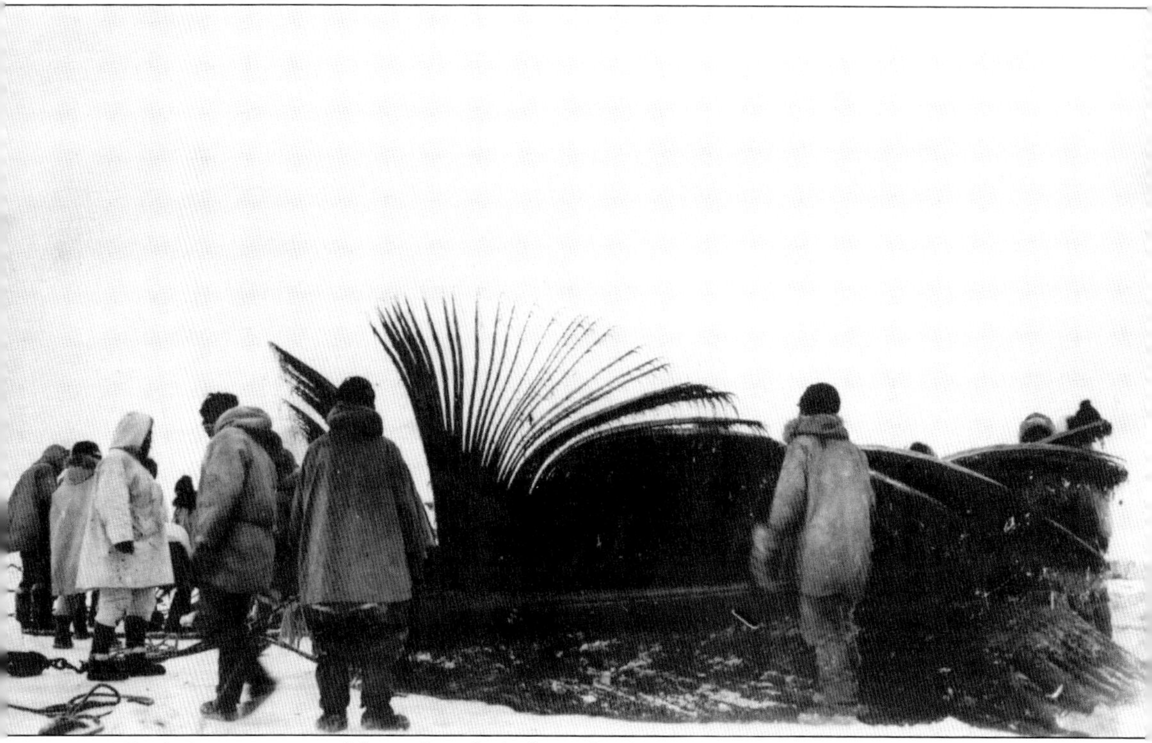

The baleen is removed from the whale from the front to the back. Baleen can be as long as 15 feet. The American whalers set up commercial trading houses in the Arctic, where they would trade with the natives for baleen and oil. So many whales were taken by the whalers that the livelihood of the natives was threatened. Many of the whaling ships, being pressed for space, would take just the baleen and oil and then discard the carcass with the meat. Occasionally, the whalers would leave a carcass for the natives, who took the meat for their villages.

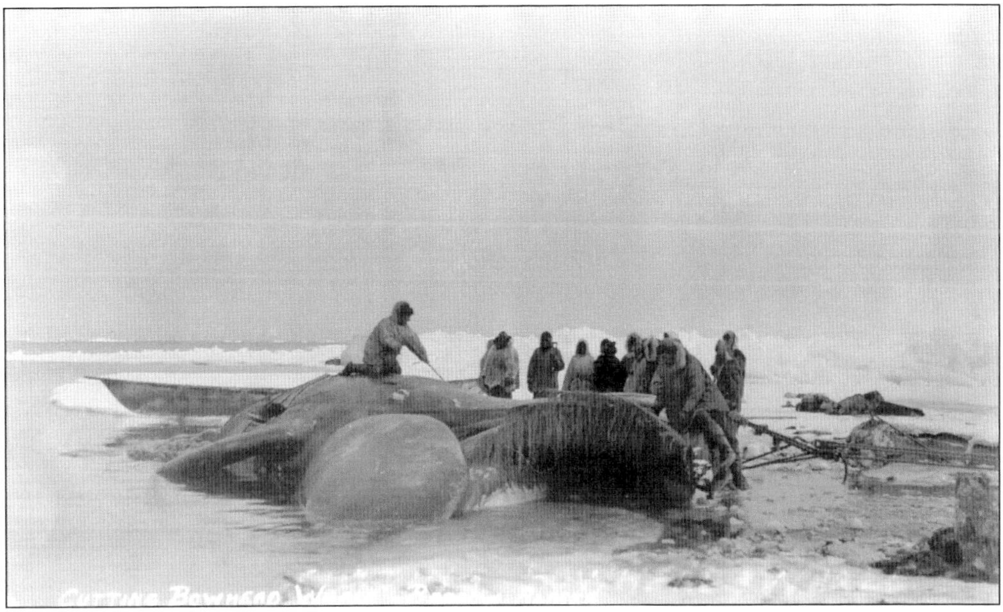

The whale has been brought to the edge of the ice to be hauled out. Until recently, a rope would be tied around the whale's tail and the entire village would come to the water's edge. The village residents would line up along the rope and work together to pull the whale out of the water to the flensing area. The whales are now pulled out by a truck or tractor, but some of the villagers still show up to view the scene.

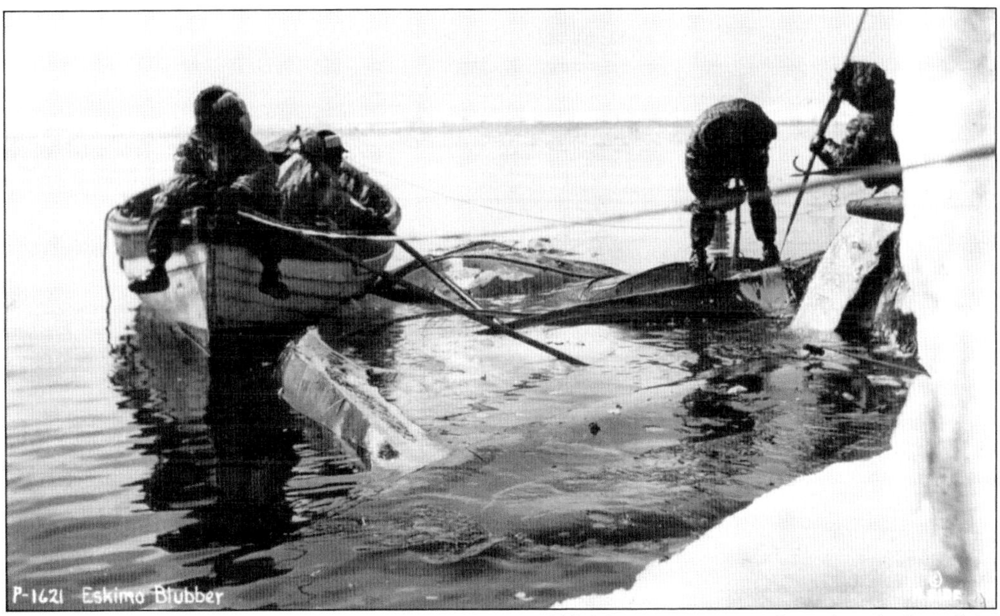

This whale is being flensed in the water as chunks of blubber are cut out of it. The whale will be rolled over in the water until all of the blubber is removed. After that, the rest of the whale—mostly meat and bones—can be towed on to the ice. Frequently, the whalers do not participate in the flensing but instead return to the water to continue whaling.

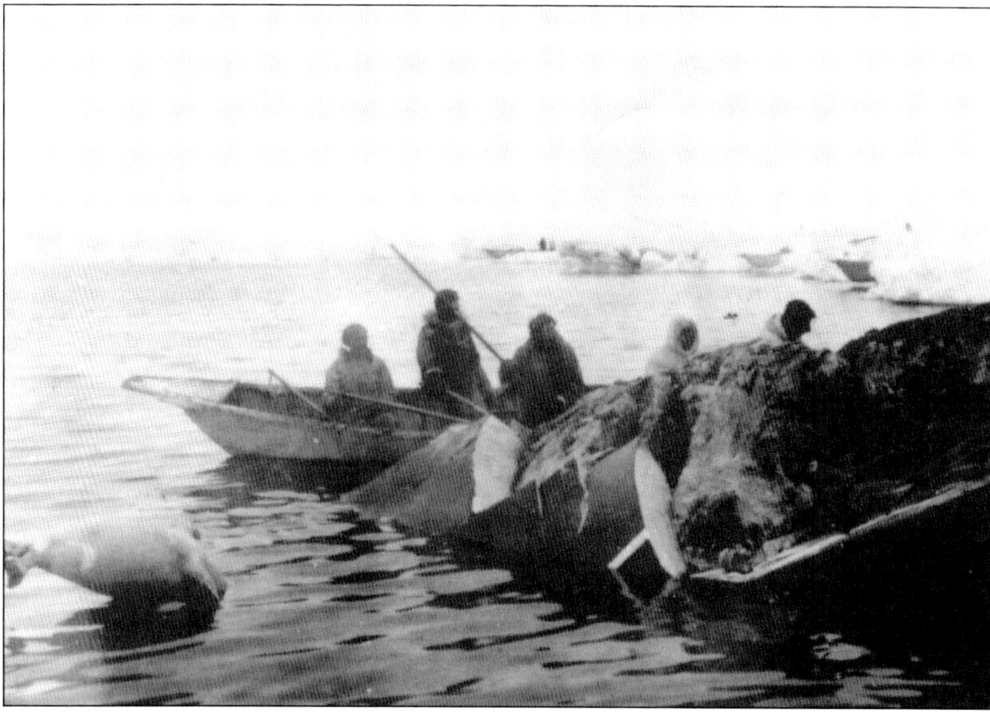

Note the thickness of the blubber shown here. This was a healthy, well-fed whale. Some whales may have little or no blubber, which may be a sign of a sick or underfed whale. A whale uses blubber to store energy for future endeavors such as migration.

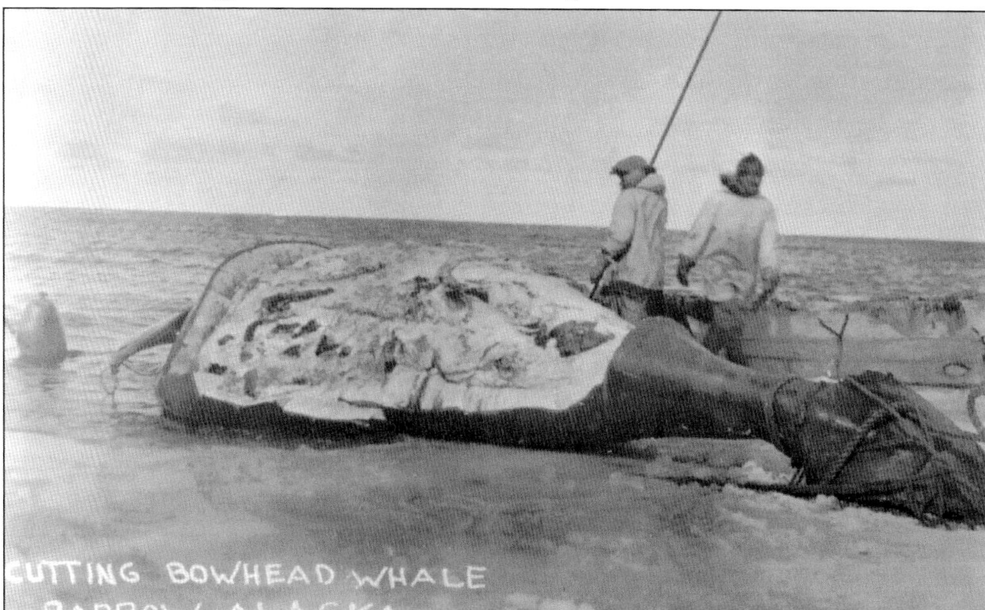

Here, the blubber has been removed from one side of the whale, which will be turned over for the removal of the blubber on the other side. Next to the whale (at left) is a sealskin float. These floats were attached to the harpoons driven into the whale. As the whale dove, the float provided buoyancy to help tire and slow down the whale.

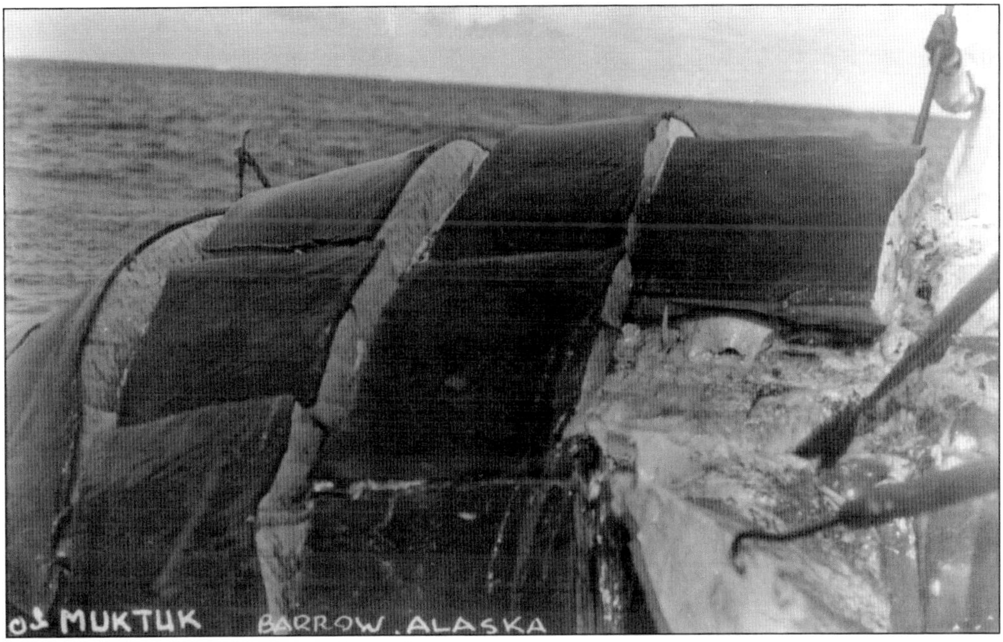

This whale has been cut vertically and horizontally in order to create pieces of blubber (or muktuk) that can be hauled away. The tools shown at right are a spade and a hook. The spade is used to separate the blubber from the body, and the hook is used to grab the piece of blubber. The piece could be pulled off the whale and dragged along the ice, where it was further cut down and distributed. Flensing on shore starts from the center of the whale; blubber is then removed in rectangular pieces. The blubber can be eaten raw or reduced to oil used for dipping whale meat at the table.

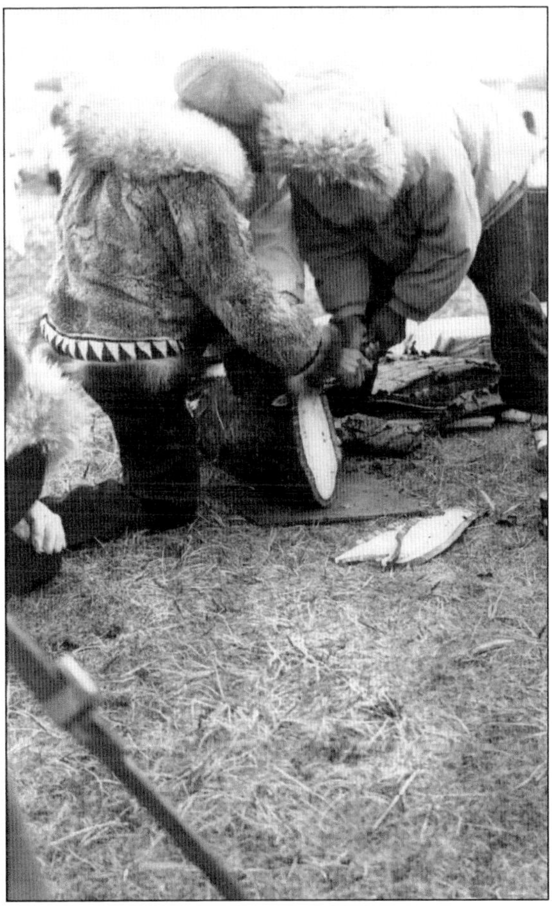

The men are slicing what appears to be the tail of a whale. The tail pieces are frozen, sliced, and then distributed at ceremonies held before and at the end of the hunting seasons.

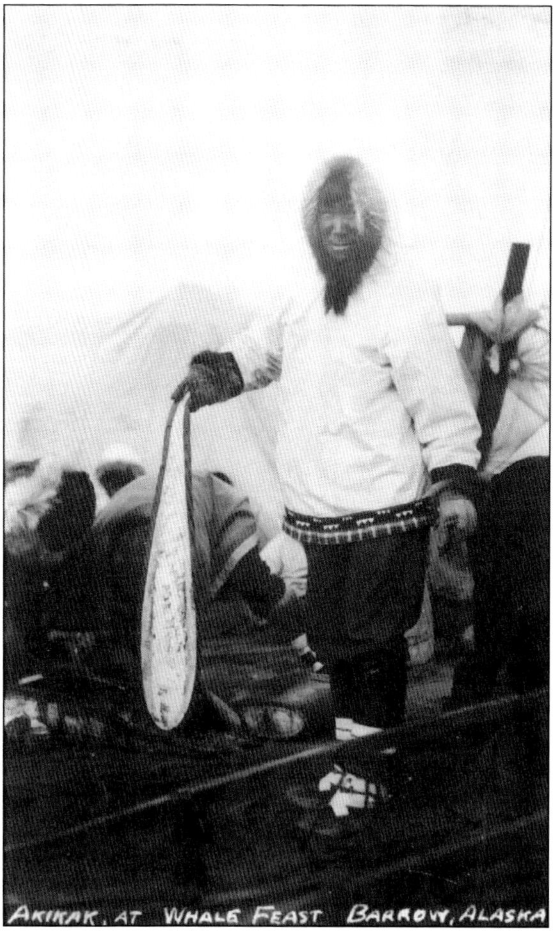

A man holds a slice of what looks like the tail of a whale. Whalers distribute meat and blubber to their own family, then to the village, and then to villagers who reside out of the area.

This dolphin or small whale is having its blubber and meat removed. Some of the cuttings will be eaten fresh, but much of it will be preserved for future needs. The native family needs to provide its own food and must gather food to support its dogs, too. Fewer dogs are needed today, because it is easier to maintain snowmobiles and ATVs than to maintain dog teams.

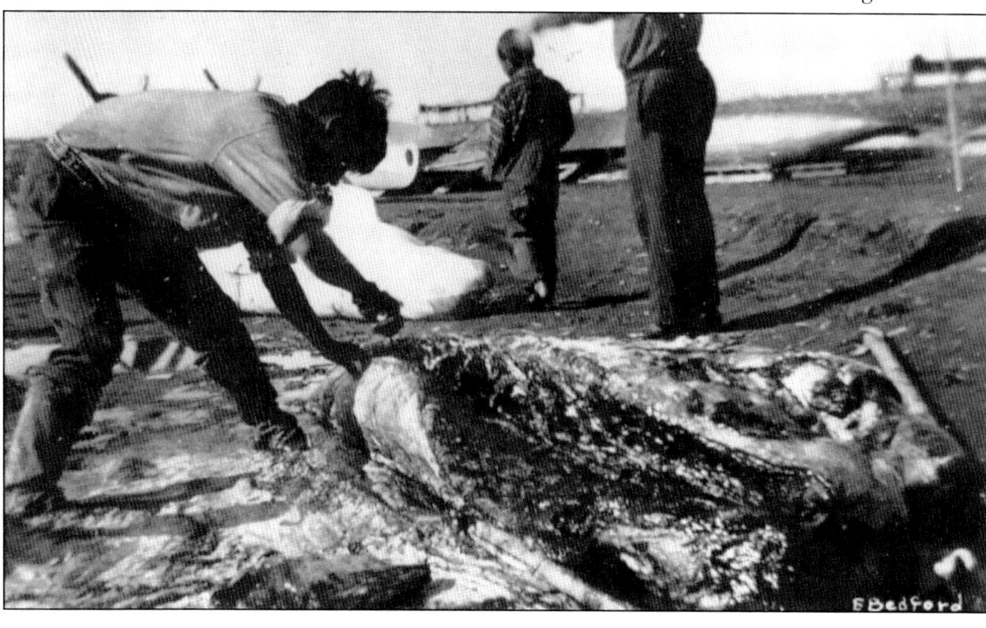

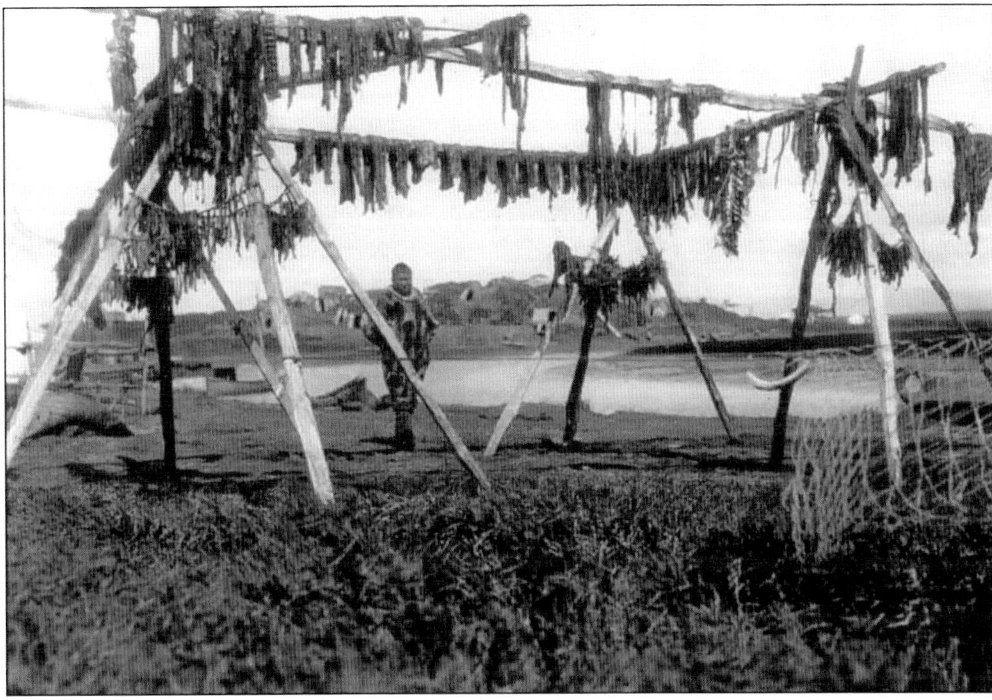

Whale and seal meat may be dried for future use. Meat will also freeze when left out in cold seasons; it is then placed in cellars. In the far north, the ground is frozen in what is called permafrost, and meat can be stored for several seasons. The meat tends to degrade slowly, so it is best to eat older meat as soon as possible.

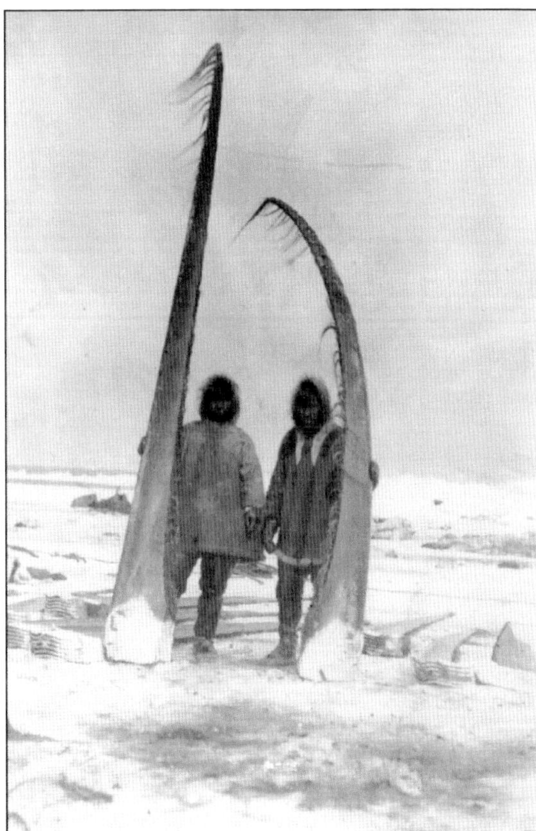

Individual baleen, which may reach 15 feet in length, looks more like trees than a body part. Baleen was once in strong demand, but around the start of the 20th century, petroleum products started replacing it in corset stays and buggy whips, as well as many other small items that could be made from plastic.

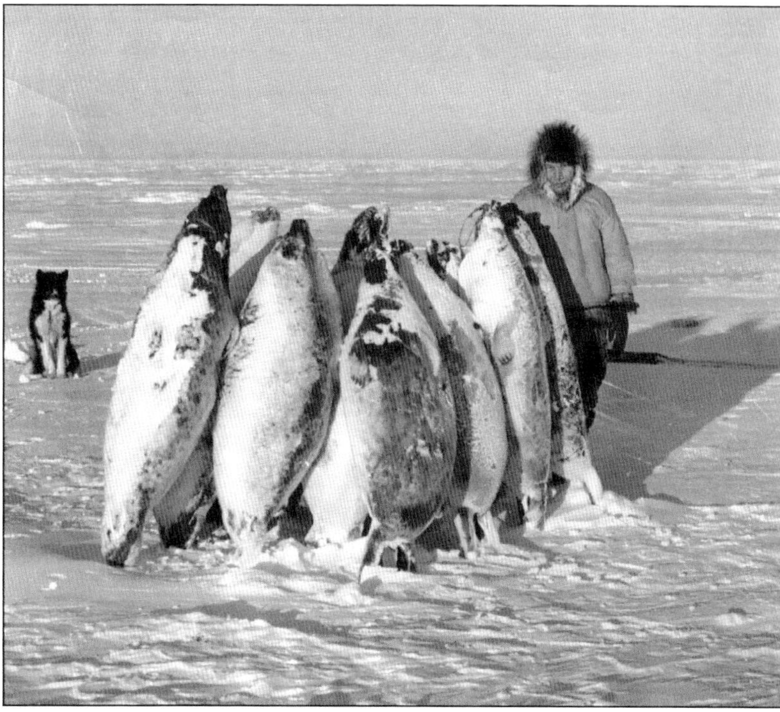

Seals are caught on the ice by the hunter. Once dead, they will freeze and thus be perfectly preserved. Frozen seals are used for dog food. The partly dried sealskins, which the captain has been saving for over a year, are tied with the ends put through a hole in the ice to soak for a week or two to soften them—they can then be used to make boat skins or floats.

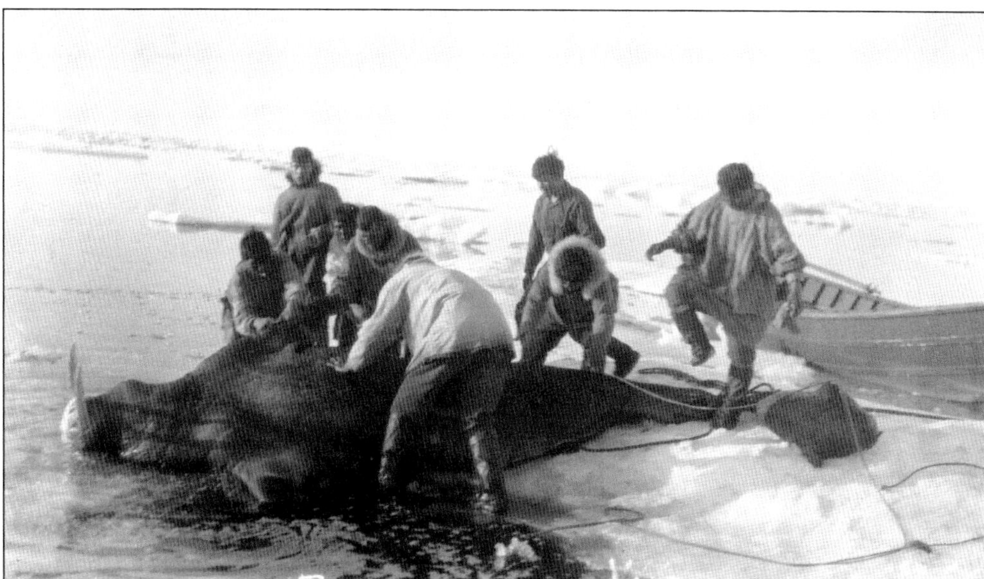

Whalers took walrus when they were available. Like whales, they provided meat and skin, but they also provided ivory. The ivory is now used to make items for the tourist industry. Walrus numbered in the tens of thousands before whalers discovered them. The whalers used to take the ivory tusks and reduce the body fat to oil, which was kept separate from the grades of whale oil.

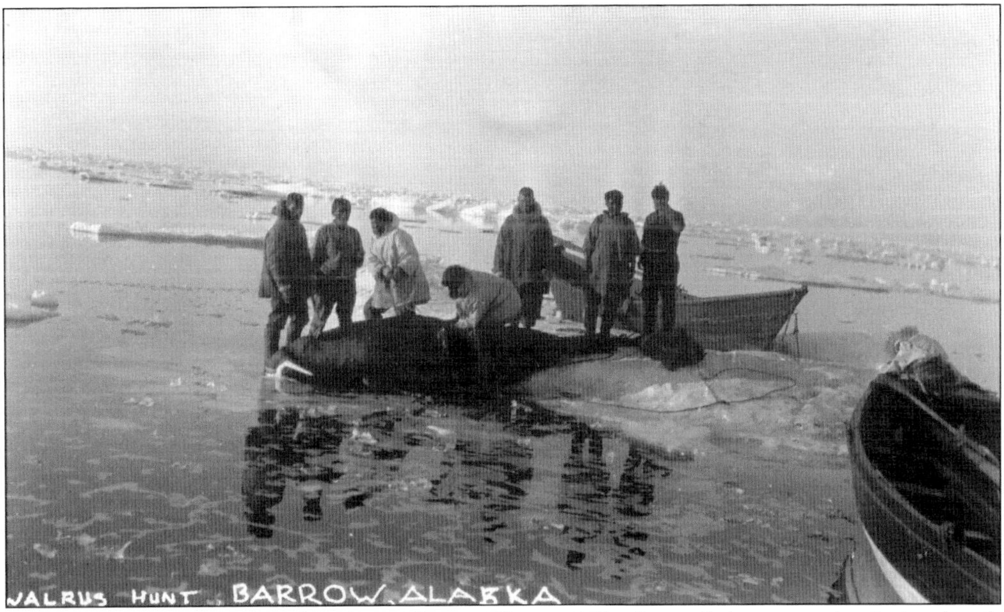

Here, a walrus is cleaned on the edge of the ice. This could be done while waiting for the bowheads to show at the start of the season. Beluga whales usually appear a few weeks before the start of the spring season.

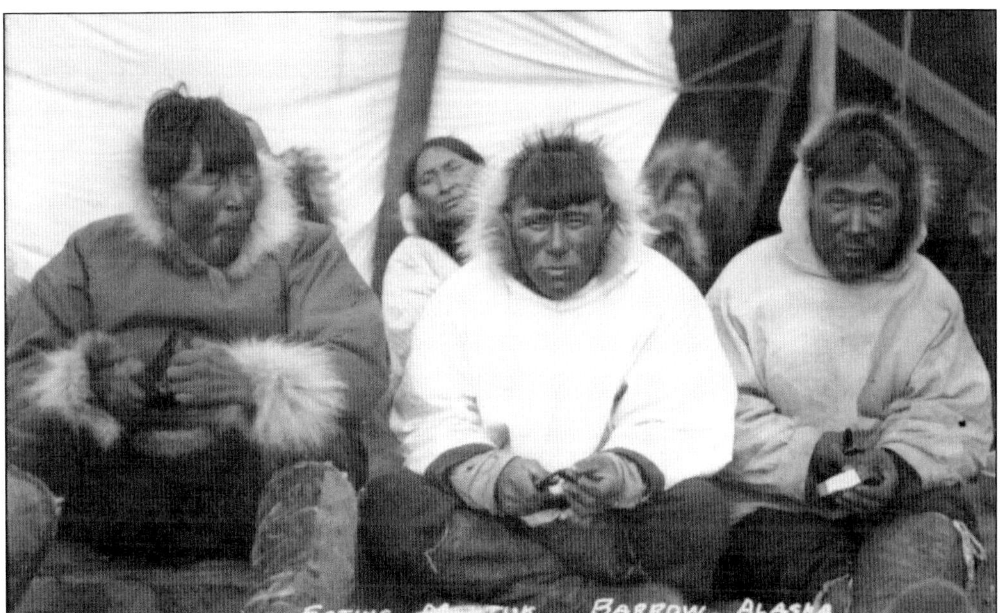

These men are sitting in a typical Inupiat cafeteria eating muktuk—a meal of frozen whale skin and blubber. The muktuk is cut into small pieces and chewed raw or sometimes dipped in seal oil.

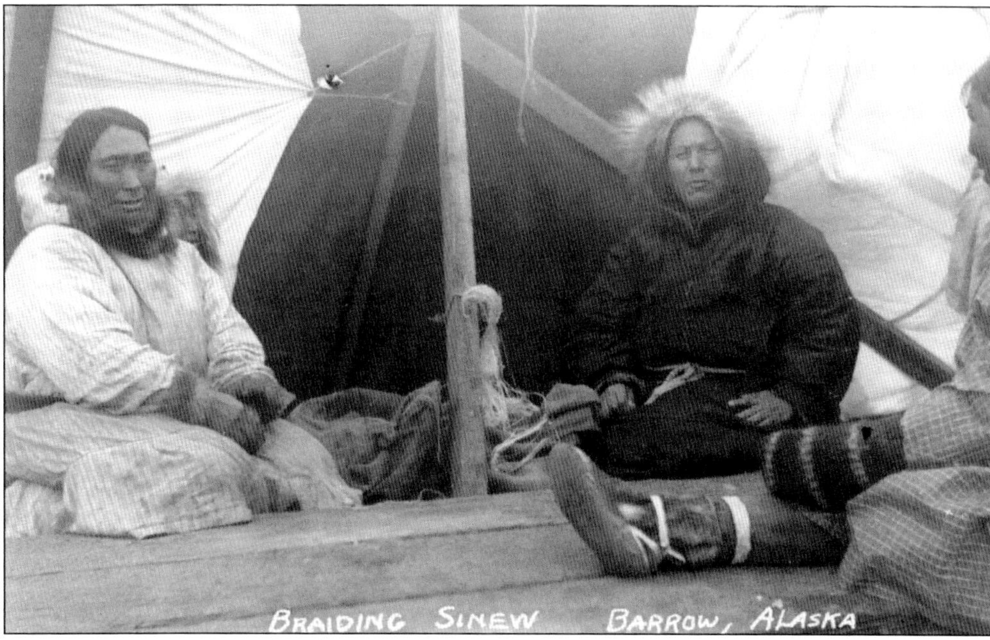

These Inupiat women are braiding sinew to make line. Sinew comes from the tendons of caribou and elk and is used to make bows, drums, and other items. Similarly, parts of the whale are used for food, tools, or ceremony. There were very few parts of the whale not put to some use.

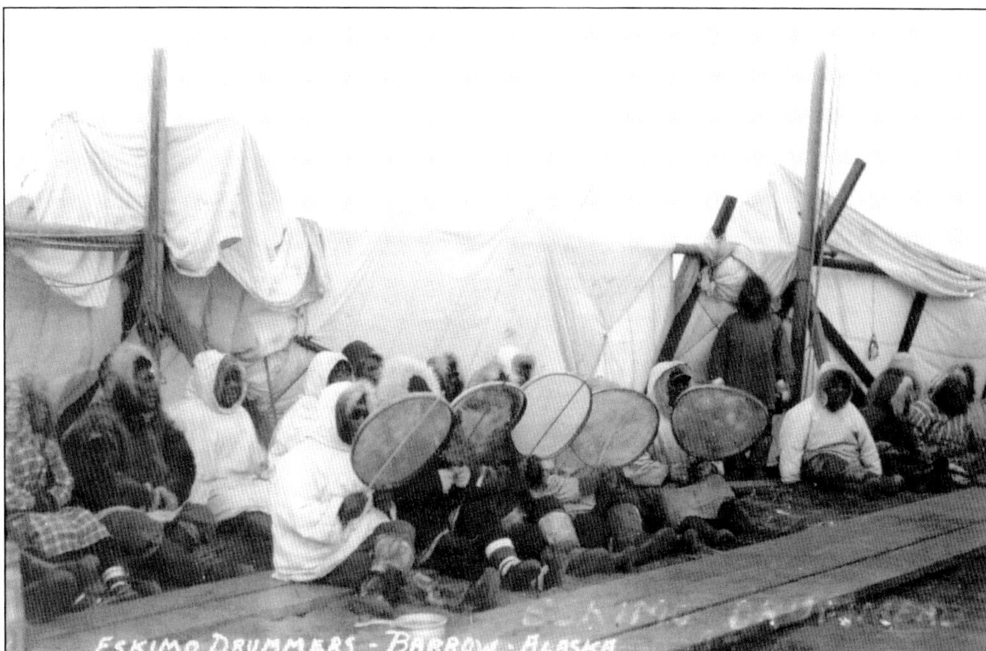

Drummers prepare to perform at the end-of-the-year ceremony. The village hosts this ceremony to honor the successful whalers. Eskimos live in family clans, with two or more clans making up a village. Whale captains typically assemble crews from their own clans, and at the end of the season, each clan hosts its own celebrations. Anyone in the village may attend the celebrations and partake in the fun and games.

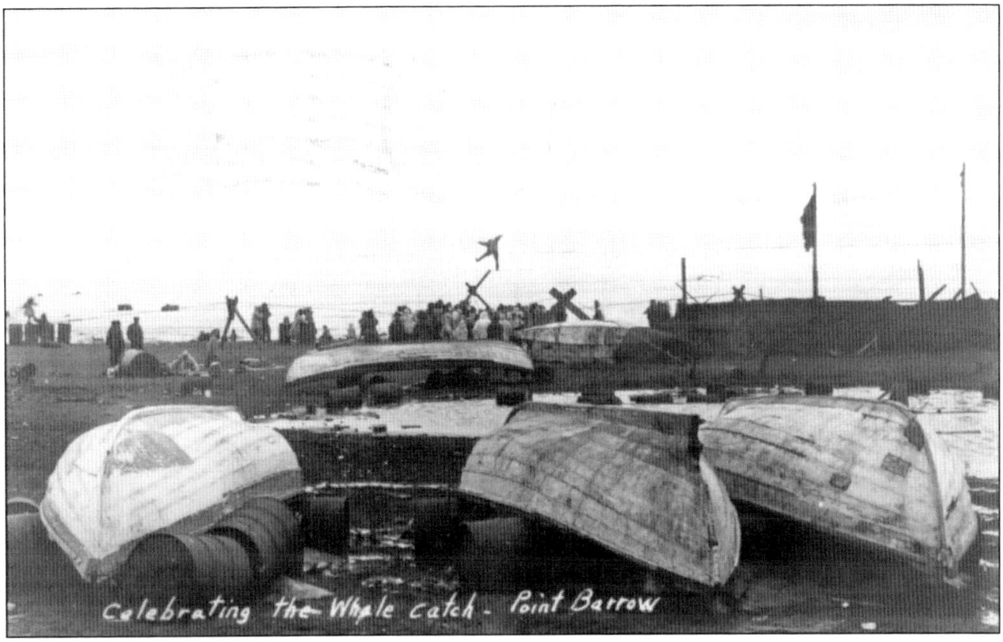

The blanket toss is a favorite event at Inupiat ceremonies. The boat captain and some of the crew are thrown, as are some village elders and prominent citizens. The blanket toss is not for children or the faint of heart, as an adult can be thrown as high as 30 feet into the air!

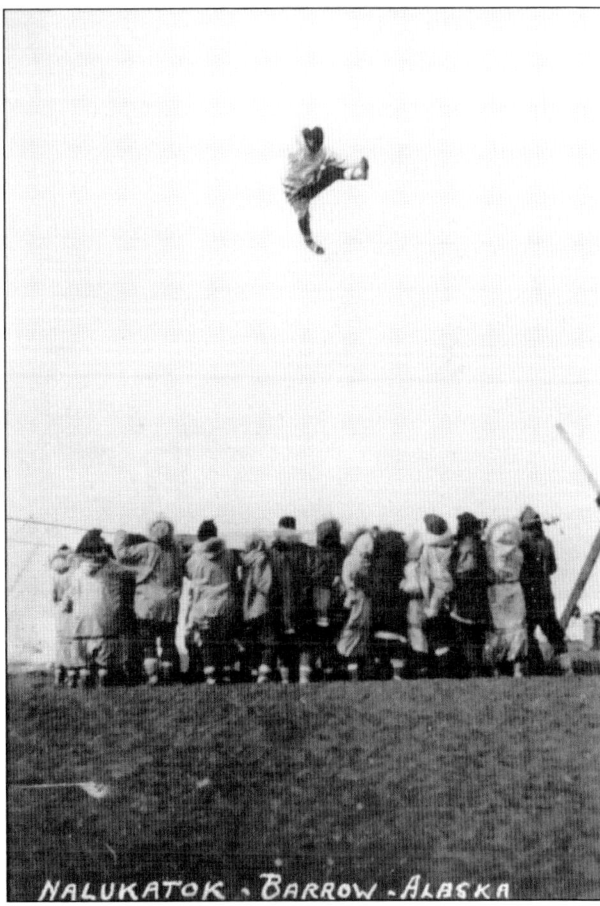

Nalukataq, the spring whaling festival of the Inupiat that takes place in June, is most famously characterized by the blanket toss. Festivals are scheduled in a way that allows people of one community to also attend festivals in others. In Barrow, which is the northernmost city in the United States, Nalukataq occurs during the third week of June.

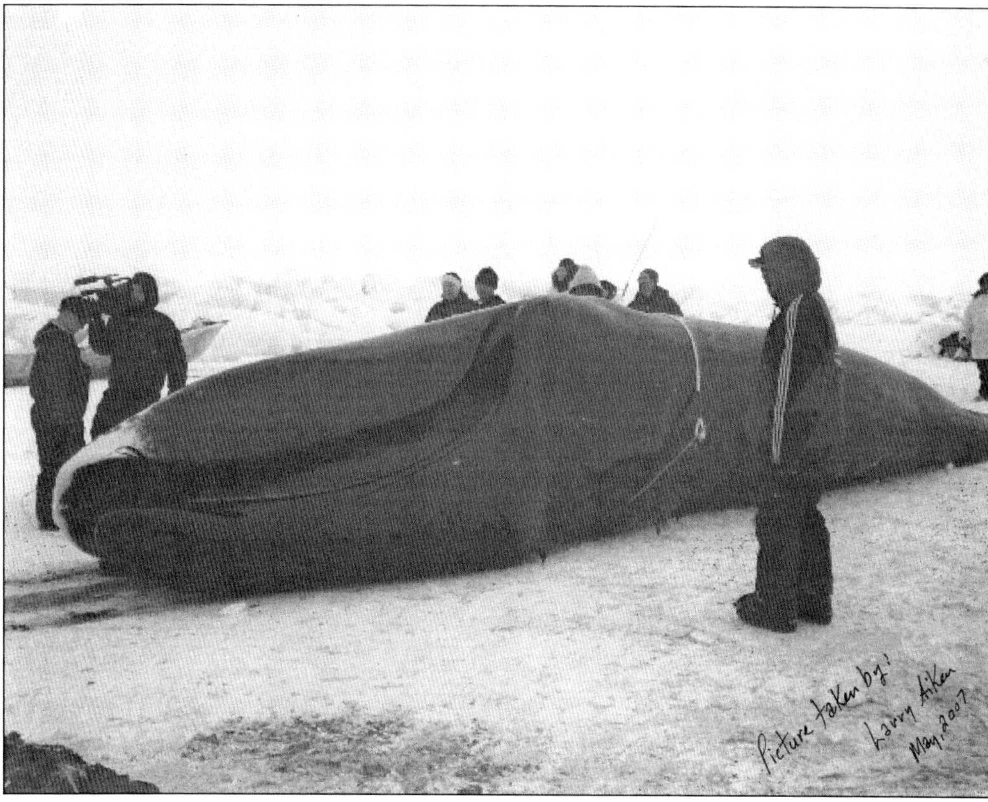

This photograph shows a bowhead whale hauled out on the ice before flensing begins. The whale was taken by Charles Hopson and his crew in the spring season of 2007, and the photograph was taken by Larry Aiken. (Courtesy of the Inupiat Heritage Center, Barrow, Alaska.)

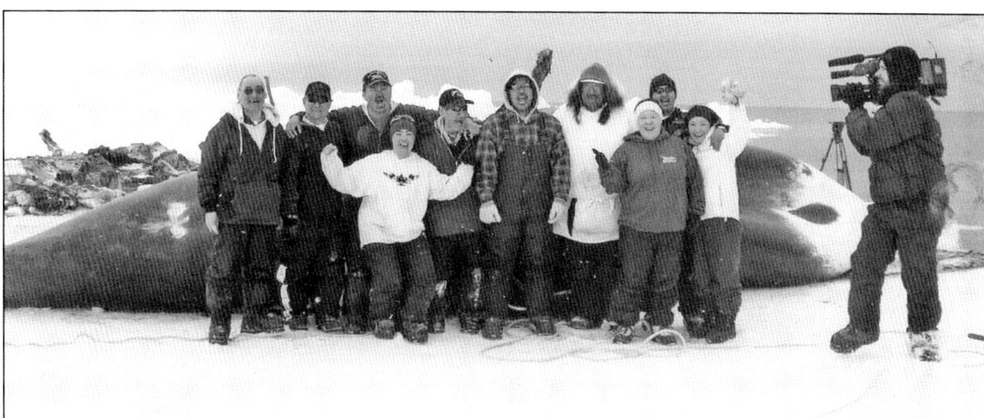

Here is a shot of the bowhead whale with the Charles Hopson crew, family, and friends celebrating and posing with their catch prior to flensing. The whale and the meat, fat, and oil taken from it are very important to the families and community of Inupiat and Inuit natives in Alaska. In the past, taking of a whale meant the village would survive the season. If the village did not take a whale, the year would be very difficult. (Courtesy of the Inupiat Heritage Center, Barrow, Alaska.)

Two

TYEE WHALING STATION

In 1907, John Leslie Barneson and Isaac Norris Hibbard, of Barneson & Hibbard Company, San Francisco, built the first shore whaling station in Alaska. It was located on Admiralty Island, one of the islands off the southeast coast. Norwegian Svend Foyn had invented the whaling station, and most of them were also constructed by Norwegians. The Tyee station, however, was strictly of American design and construction. It was named the Tyee Whaling Company after Tyee Cove, where it was located. The manager of the station was indeed Norwegian, as were the harpooners who would turn out to have a monopoly on this position at all whaling stations.

The contract for the "killer" boat was awarded to the prestigious Moran Shipbuilding Company of Seattle, Washington. In 1907, the *Tyee Junior* was completed by Moran's (another ship already bore the name *Tyee*). The *Tyee Junior* was the first North America–built steam-powered whale chaser. At 71 tons, the boat had a length of 97.9 feet long, a beam of by 17.7 feet, and a draft of 11.8 feet. It was not a copy of patented Norwegian models, being eight feet longer with a shallower draft. The *Tyee Junior* was equipped with two coal-fired power plants instead of the standard oil-fired engine of Norwegian use.

The whale catcher *Tyee Junior* was an entirely new design with electric lights. Not surprisingly, it proved very expensive, at a cost of $65,000, whereas the customary price for Norwegian vessels was $24,000.

Upon completion in 1907, notes Webb in *On the Northwest*, the *Tyee Junior* steamed to Victoria, British Columbia, to meet eight Norwegian whale men. "These men had been working at Sechart prior to completion of the Admiralty Island plant. During the visit, a party of local businessmen enjoyed a ride aboard the new steamer, and among them was William Grant, president of the Pacific Whaling Company. . . . After the inspection the *Tyee Junior* steamed north to the new station at Murder Cove." The *Tyee Junior* proved itself the fastest whale chaser on the coast with a top speed of 13 knots.

Two small towns, Juneau and Skagway, were not far from Admiralty Island. These towns, according to Tønnessen and Johnsen's *History of Modern Whaling*, became "famous and notorious as starting points for the gold-diggers' adventurous trek across the mountain passes to the Klondyke." This began in the 1890s, as "tens of thousands of men were drawn there," and shipping and whaling interests had to put in great effort in order to retain crews.

Commercial whale catching commenced in the summer of 1907, but as early as 1909, humans began to realize that the whales were becoming shyer and more inclined to keep to the open ocean. "To overcome this difficulty," write Tønnessen and Johnsen, "a remarkable form of floating factory was rigged up . . . on two barges, which were towed out and anchored near the town of Sitka on Baranof Island." The oil-making machinery was placed aboard a large barge called the *Diamondhead*, and during the first half of the season, whaling commenced around the Shumagin Islands and Kodiak Island, where whalers utilized the barge to cut up the catch and try the oil. The floating station returned to Admiralty Island in the fall, at the close of the season.

In 1910–1911, the Tyee Whaling Company operated from Whale Bay in southeast Alaska with the *Tyee Junior*, the tugs *Fearless* (which did not whale in 1911) and *Resolute*, and a second catcher—the *Lizzie S. Sorenson*. The *Lizzie S. Sorenson* earned a place in whaling history on May 10, 1910, when it became the victim in one of the few authenticated cases of a boat being sunk by a whale. An 84-ton schooner, the *Sorenson* had been purchased the previous year and retrofitted with a gasoline engine. The story says that a harpoon had found its mark in a large blue whale but only wounded it. The whale careened off in a wide arc and returned at full speed, ramming the side of the boat with its head. The whale repeated this maneuver several times over, finally sinking the *Sorenson* near Cape Addington. Crew members who survived the attack believed that the whale had acted deliberately—aiming to sink the vessel—rather than mindlessly thrashing about in reaction the harpoon's pain. The next day, another ship discovered that a dead blue whale had surfaced near the location of the event.

In 1911, the Tyee Whaling Company altered its original plan of operation and became engaged only in the production of whale oil. In spite of the reduction in the fleet, the company enjoyed a good year with a high percentage of blue whales in the catch.

On November 3, 1912, the Tyee Whaling Company schooner *Allen A.* arrived in San Francisco from the company's station at Whale Bay with 5,500 gallons of whale oil, 25,500 pounds of humpback whalebone, and 65 tons of whale fertilizer. The schooner had a very hard trip down the coast, taking 32 days for the passage, during which several sails were carried away and the boom was broken. The *Allen A.* had not been away from Alaska in two seasons (since 1909) when it went out on a trading and fishing cruise off the Alaska coast. The following year, 1910, it went aground near Kodiak Island and was acquired by Tyee after being taken off; the company then operated it in connection with its whaling and trading business.

While the rapid development of this industry on the Pacific Ocean resulted in a considerable increase in the products over any previous year, 1912 did not meet expectations. The catch was smaller than anticipated, and the year's whales were thin and yielded much less oil than usual. Bad weather also played a part in curtailing the operations of some whaling concerns. Two other stations started in 1912, taking 300 whales while the Tyee Company took only 61.

The Tyee Whaling Company ceased operations before the 1913 season. All the company's machinery from the reducing plant was brought to Seattle in late June on the oil barge *Fresno*, which—together with the flensing scow *Sperm*—was towed by the tug *Goliah*. In addition to the machinery, the barge held 200 barrels of oil. The whaling equipment of the *Sperm* attracted considerable attention on the Seattle waterfront.

Barneson sold the cut-down, American-built barge *Fresno* (it had originally been a bark) to the North Pacific Sea Products Company, which used it to transport men and equipment to and from Akutan. In 1913, Tyee Whaling Company was also negotiating with a San Francisco company for the sale of the whaler *Tyee Junior* and the barge *Diamondhead*. The equipment was in Seattle, where it was brought after the company had suspended operations.

By January 1916, Tyee's 10-acre parcel of land on Admiralty Island, with buildings "suitable for a salmon cannery," according to an article in *Pacific Fisherman* magazine, were available for sale or lease.

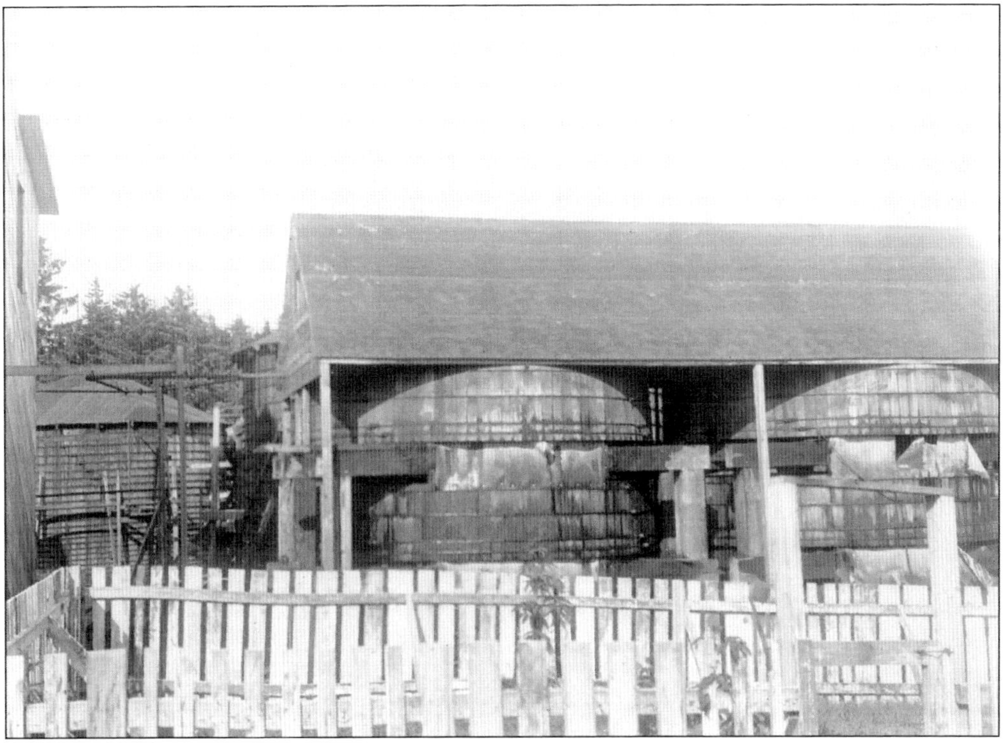

John Nathan Cobb, a conservationist who became a field agent for the US Bureau of Fisheries, took this photograph of the Tyee Whaling Company oil tanks on January 28, 1911. (Courtesy of University of Washington Libraries, Special Collections, UW Cobb 3408.)

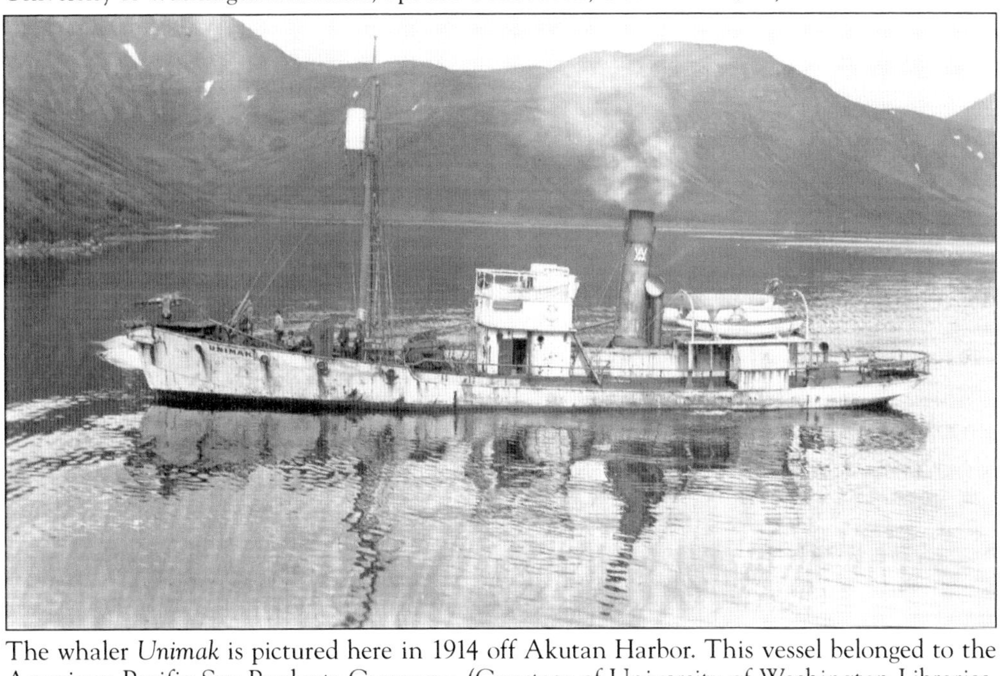

The whaler *Unimak* is pictured here in 1914 off Akutan Harbor. This vessel belonged to the American Pacific Sea Products Company. (Courtesy of University of Washington Libraries, Special Collections, UW 14292.)

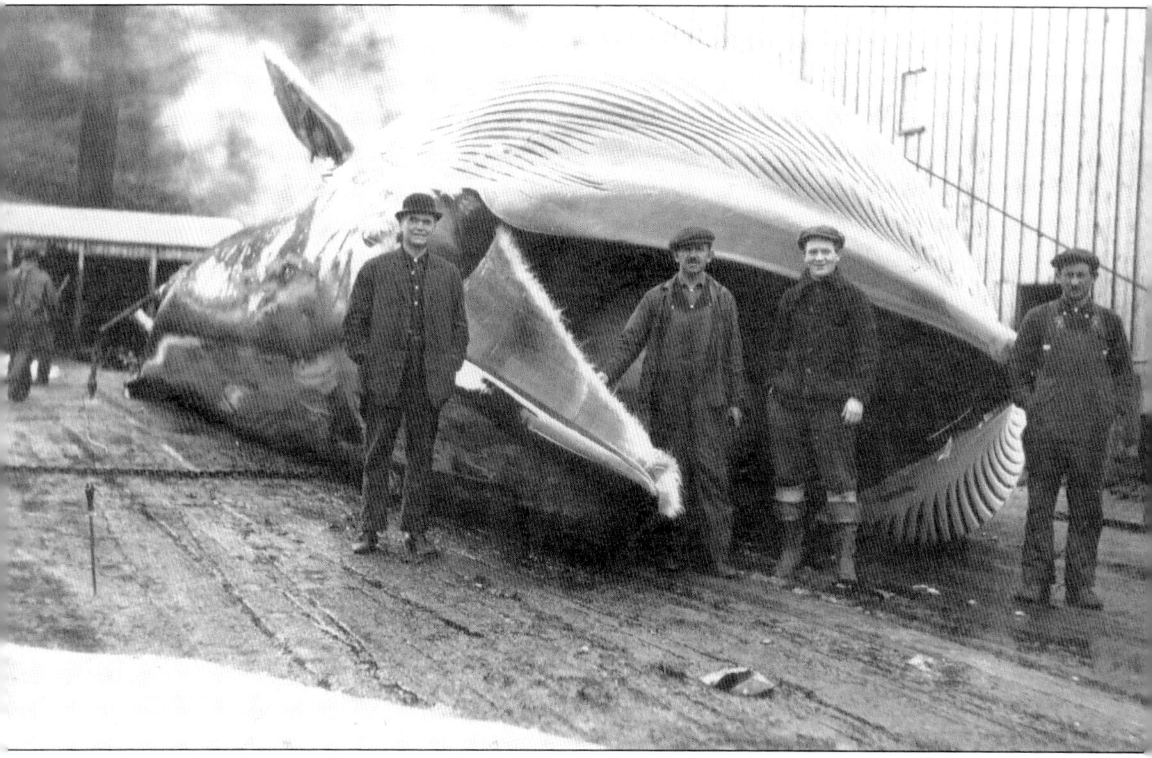

In another photograph taken by John Nathan Cobb, a group of men poses in front of a dead whale at Akutan Harbor in 1914. The men worked for the American Pacific Sea Products Company. (Courtesy of University of Washington Libraries, Special Collections, UW 14291.)

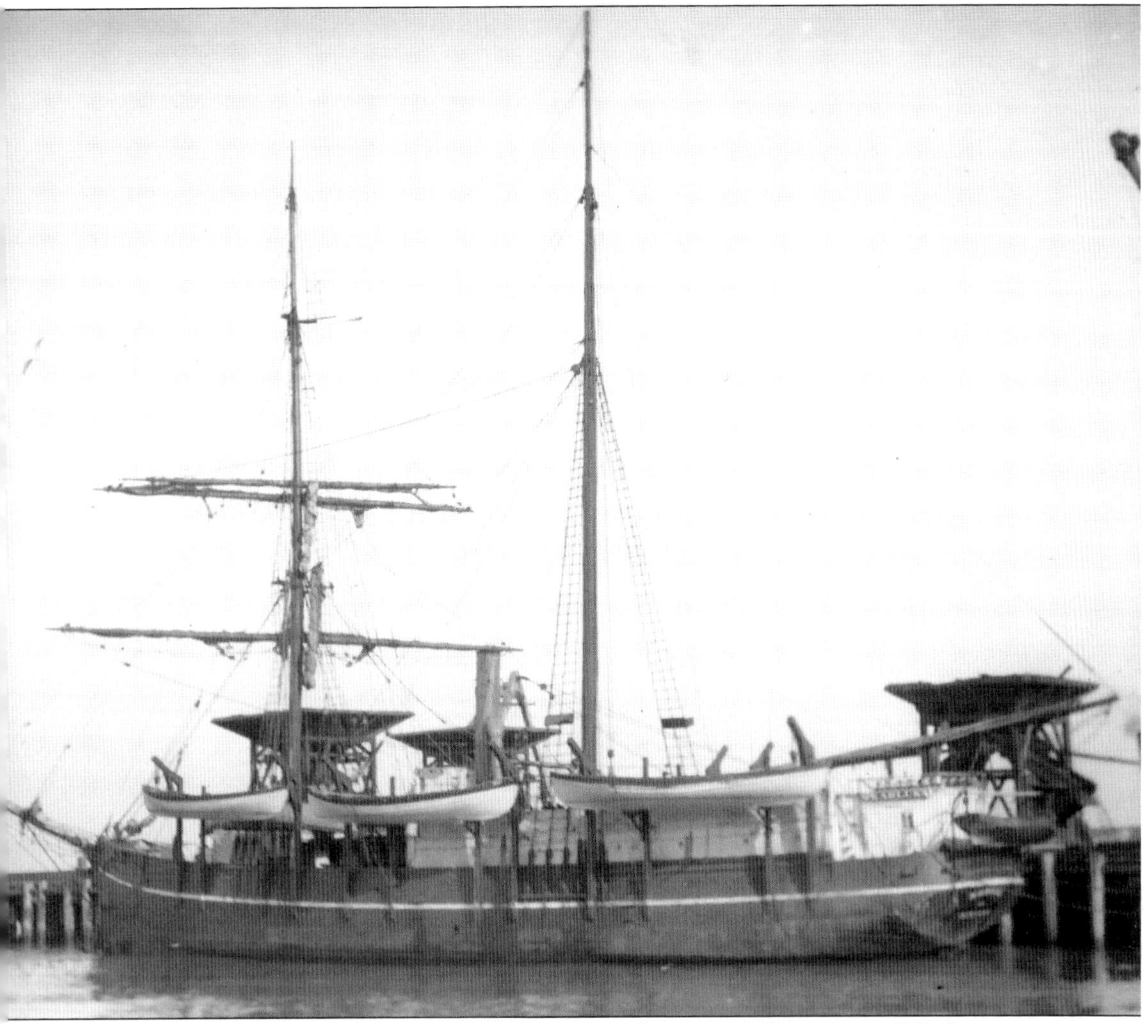

The whaling steamer *Karluk* is shown here at dock in 1913. (Courtesy of University of Washington Libraries, Special Collections, UW 14293.)

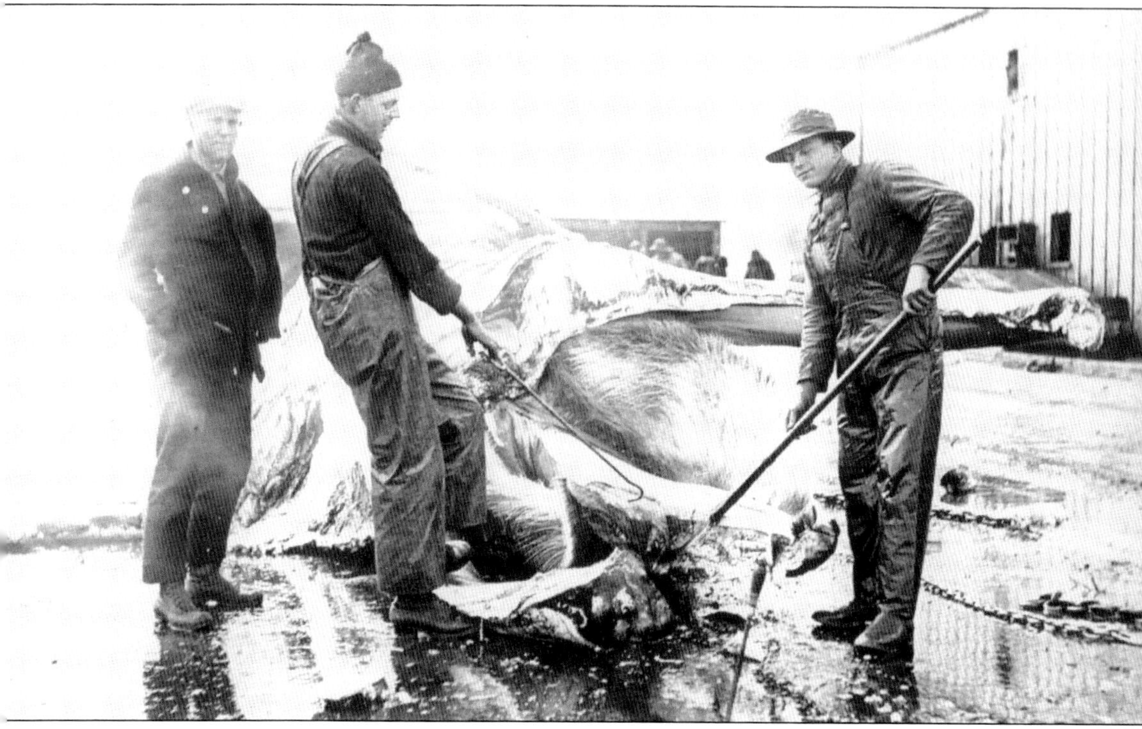

These three men are flensing a whale at the Tyee Company whaling station in 1915. John N. Cobb took this photograph. (Courtesy of University of Washington Libraries, Special Collections, UW 33493.)

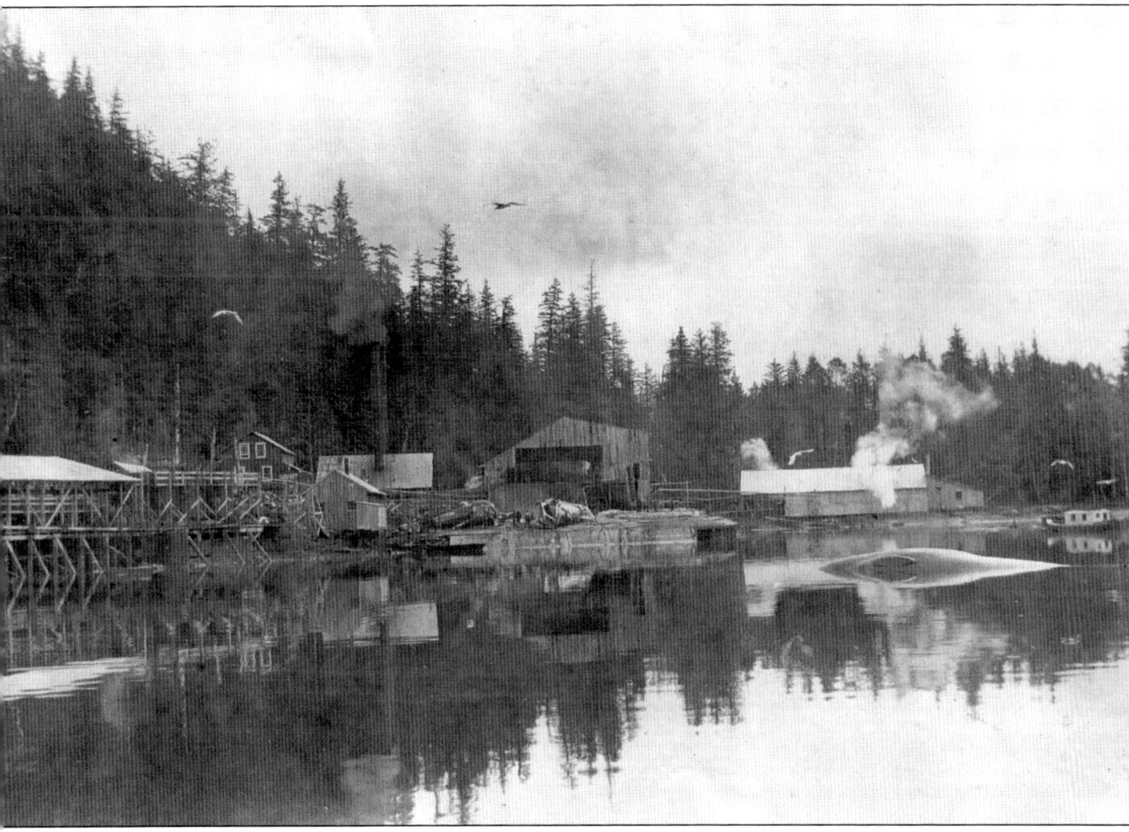

The Tyee Whaling Company established the first whaling station in Alaska at Murder Cove on Admiralty Island in 1907, and it operated until 1913. The first modern Norwegian shore whaling station in the United States opened in 1912. Whale carcasses are visible in the center of this image, straight ahead near the flensing platform. The Norwegians invented the modern whale catcher and designed the whaling station. Most of the modern whaling stations around the world up to this point and after were of the Norwegian design. The Norwegian design was able to make the most complete use of the whale.

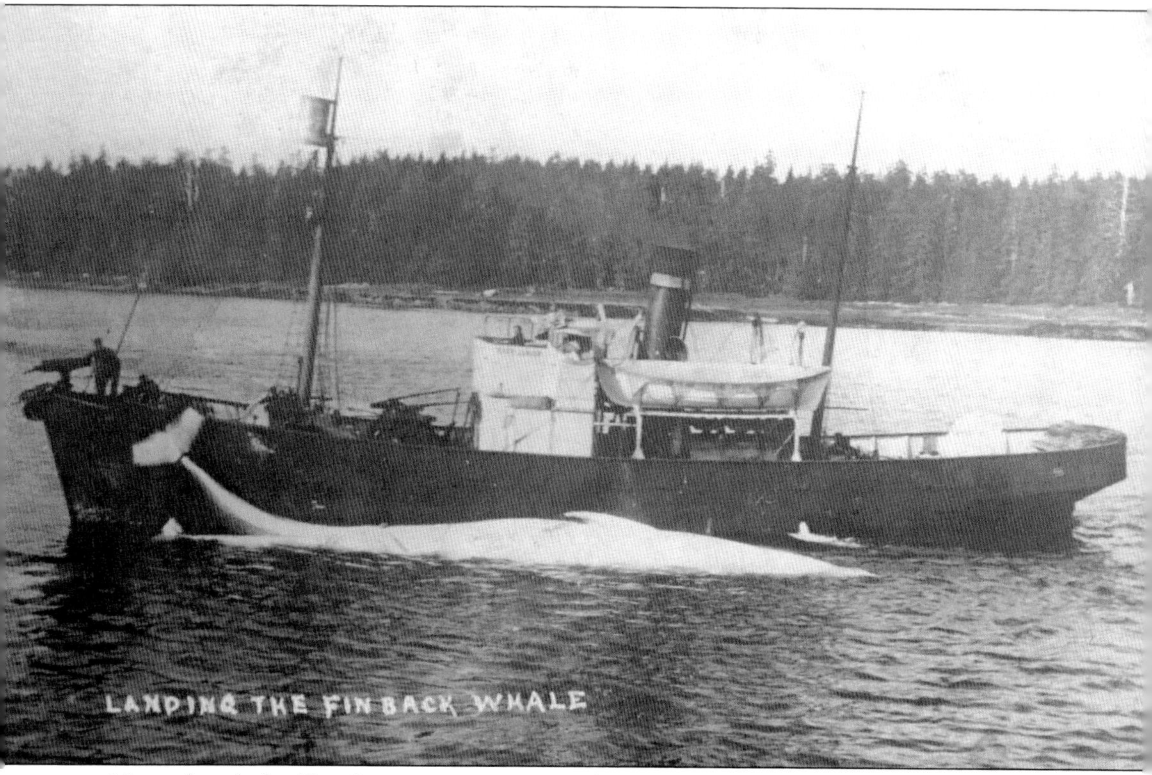

Here, the whaler *Tyee Junior* is coming into the Tyee station with a finback whale. The Tyee Whaling Company established its operations using the first American-made steel whaler—the 97-foot *Tyee Junior*—which was equipped with a harpoon gun on the bow. Consequent to a declining whale population, the company closed its operations in 1913.

Three

BELUGA WHALING

The five stocks of beluga whales found in Alaska are the Beaufort, Chukchi, Bering, Bristol Bay, and Cook Inlet. The Cook Inlet stock, having been whaled for some years, was not as healthy as the others. There are no records, but it was known that the Beluga whale, especially in Cook's Inlet waters, had been taken for years. Other whaling consisted mostly of some local fishing, most of it being for subsistence by natives all along Alaska's coast. The first commercial taking of the Beluga whale on record was by the Beluga Whaling Company in 1915.

Joseph A. MaGill, the company's owner-operator, had fish traps and canneries as businesses operating in the area of Cook's Inlet when he started the Beluga Whaling Company. When he started in 1915, he had 12 employees who rendered oil from the blubber and carefully removed the white skins to be made into leather for soft leather gloves.

In 1916, MaGill was endeavoring to interest potential stockholders in the matter of establishing a plant in Alaska for catching and utilizing belugas, also called white whales. According to MaGill, the chief branch of the "industry will be that of marketing the hides of the beluga, or white whale. Countless number of this species of whale abound in Cook Inlet and in the waters of Bering Sea." The Beluga Whaling Company was incorporated at Anchorage under the laws of the territory of Alaska with a capital stock of $25,000 in September 1916. The officers and stockholders of the company included president J.A. MaGill, secretary and treasurer Claude C. Coleman, and other stockholders A.A. Shonbeck, J.T. Westermann, H.R. Warren, and James O'Reilly.

The belugas were said to be in "enormous numbers in many sections of Alaska," according to the 1921 article "A Remarkable Whaling Development" in *Pacific Fisherman*, and were immensely destructive of salmon. They averaged 12 to 14 feet in length and about a ton in weight and produced 70 to 100 square feet of hide from which is made one of the strongest leathers known, in addition to three grades of oil from the blubber, jaw, and head, the latter being practically unfreezable.

Despite being seen as destructive of the salmon and commercially desirable for their products, very little had been done to harvest them, as belugas are very wary animals. The hides average from 50 to 100 pounds each and measure from 7 to 25 feet in length. They brought a price of 36¢ per pound, and there was always a demand because they made a leather as close to waterproof as possible. The whales were caught with nets, since belugas ran on shore when attacked. To harpoon the whales damages the hide, and damaged hides do not command a good price. The belugas, however, were plentiful and could be captured if the proper equipment was at hand. The carcass of the white whale was valuable as a fertilizer, and the oil had a commercial value.

In December 1920, Vol. 18 the *Pacific Fisherman* newsletter reported that J.A. MaGill & Co. operated a whaling station at the mouth of the Beluga River, "having succeeded the Beluga Whaling Company of Chicago. The company had a fair season, having brought out a good lot of beluga hides and oil." The company planned to see a significant increase in operations for 1921, when it would have a plant for the reduction of whale meat and offal. "All the company's products were handled through the Seattle-based F.H. Knowlton Company."

MaGill appeared to have solved the problem, at least in terms of favorable conditions. With the big tides of Cook Inlet, the lower part of the Beluga River was filled with water at high tide but practically bare when the tide went out, and a system was found to trap the belugas in the river at high water.

MaGill had been using a net of extra heavy cotton seine twine; it was 1,550 feet long with cork and lead lines 1.5 inches in diameter. "The lead line was lashed to stakes driven near the mouth of the river," the 1921 *Pacific Fisherman* article reported, with a 2.5-inch hose "attached to the corkline connected to a pipeline that, in turn, ran to an [air] compressor and tank." The tank was installed far enough away from the beach it was out of sight and sound from the water. As the tide came in, the net lay on the bottom, so nothing unusual was visible to the whales. The belugas then followed the salmon into the river, and when they swam above the net, the hose was inflated, rising to the surface and making a complete corral around the mouth of the stream. Keg buoys were then attached at intervals to prevent the net from sinking again.

The MaGill plant had three large rendering tanks for the blubber oil, a high-grade light oil with less free fatty acid than ordinary whale oil. The jaw oil was rendered by the heat of the sun and used as a fine watch oil. MaGill said there was a large quantity of fine edible meat on a beluga, but no effort was made to market it anywhere except locally, where it was used for dog food.

The Beluga Whaling Company operated from 1917 until 1921 at the Beluga River in Upper Cook Inlet and harvested 151 belugas (9, 42, none, and 100 between 1917 and 1920, respectively) before going bankrupt.

Capt. Bernhard Nygren, who for some years was engaged in catching beluga whales in the vicinity of Cape Nome, was president of the Arctic Sea Products Company, which in June 1917 was based in Seattle with a capital of $51,000. The other incorporators were H.J. Gangmark, Thomas R. Finch, O.L. Anderson, and Joseph P. Totten.

According to the 1916 *Report of the Commissioner of Fisheries*, Capt. Ben Nygren and S. Torkensen had a small station on Cape Nome, about 12 miles from the city of Nome. Before 1917, they caught Beluga whales only for dog food, but under war conditions, they found a demand for oil and hides, which produced extremely strong leather. Captain Nygren also considered the shipment of the whale meat to Seattle, having explored the possibility with shipping companies.

The output of hides in 1919 was shipped to Europe, as a market for the hides in the United States was lacking. The new company was preparing to conduct the work on a somewhat larger scale than the previous year. In 1920, the company was recorded as having taken 136 whales. In 1920, the Arctic Sea Products Company was stricken from the state records list for failure to pay an annual license fee, and nothing more about the company can be found.

J.H. Young was hunting beluga whales at his Golovin station in 1919–1920, getting about 75. "There was a great run of belugas in Norton Sound," Young reported to *Pacific Fisherman* in 1920, "to where they could often be seen jumping for many miles, destroying vast quantities of salmon and herring." Young brought down the hides and oil and found a good market for the meat, which was wanted for dog food. He also brought some walrus hides and oil. J.H. Young and Company was reported as making plans for business operations in 1920 and 1921, but no results have been found, and it is assumed the company went out of business.

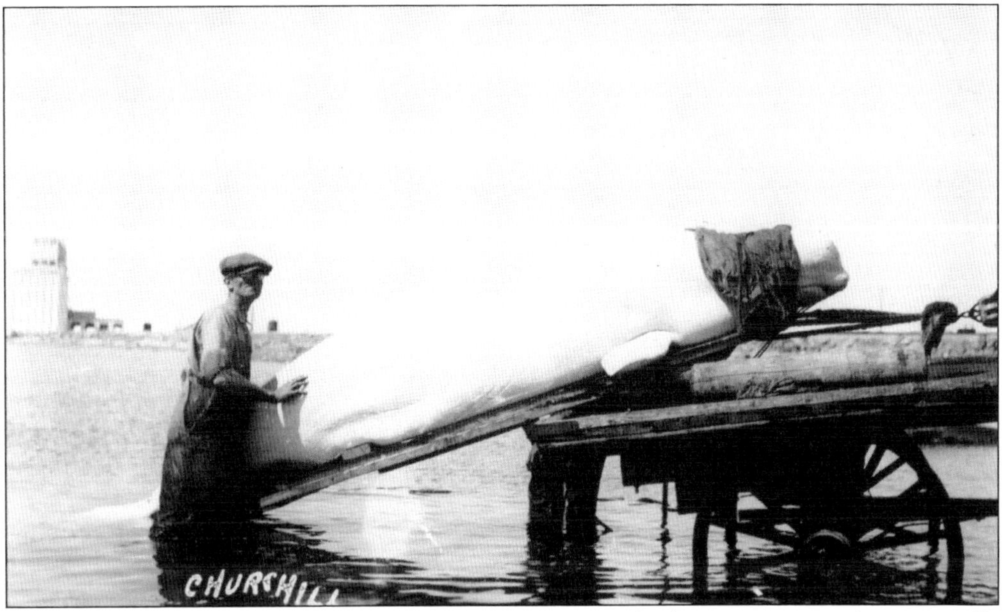

Beluga, or white whales, are distributed widely in Arctic and subarctic waters and are generally associated with areas seasonally covered by sea ice. Beluga whales are found in Cook Inlet, Bristol Bay, the eastern Bering Sea, the eastern Chukchi Sea, and the Beaufort Sea. The Cook Inlet population is the most isolated of the five and resides year-round in the waters of Cook Inlet.

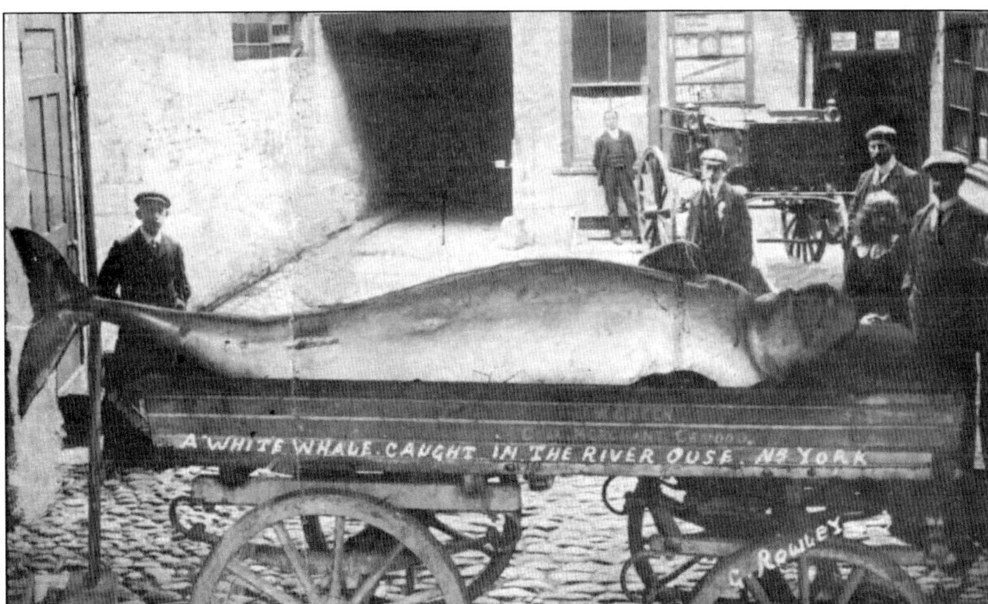

Belugas average 12 to 14 feet in length and about one ton in weight; they produce 70 to 100 square feet of hide from which very strong leathers can be produced. "Beluga" is derived from the Russian word *belukha*, meaning "white."

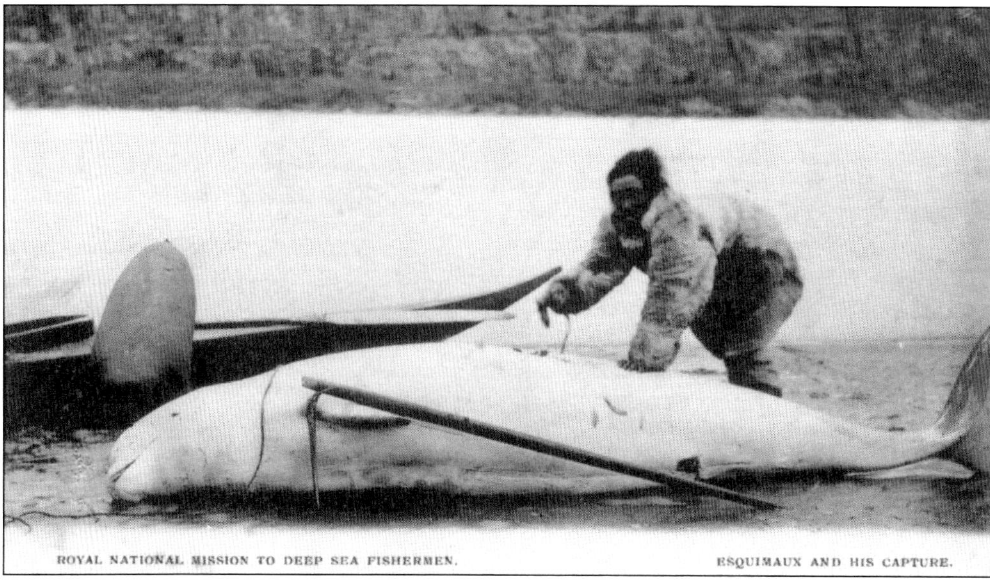

The harpoon has a point on the end that toggles when the head enters the whale, and the shaft of the harpoon drags behind, as shown. Floats are tied to the line, which slows down the whale by providing drag. When the hunter draws up to the whale, he will take out a lance and stab the whale, attempting to hit a vital spot. Once captured, the whale will be towed by the tail or with a rope through the mouth.

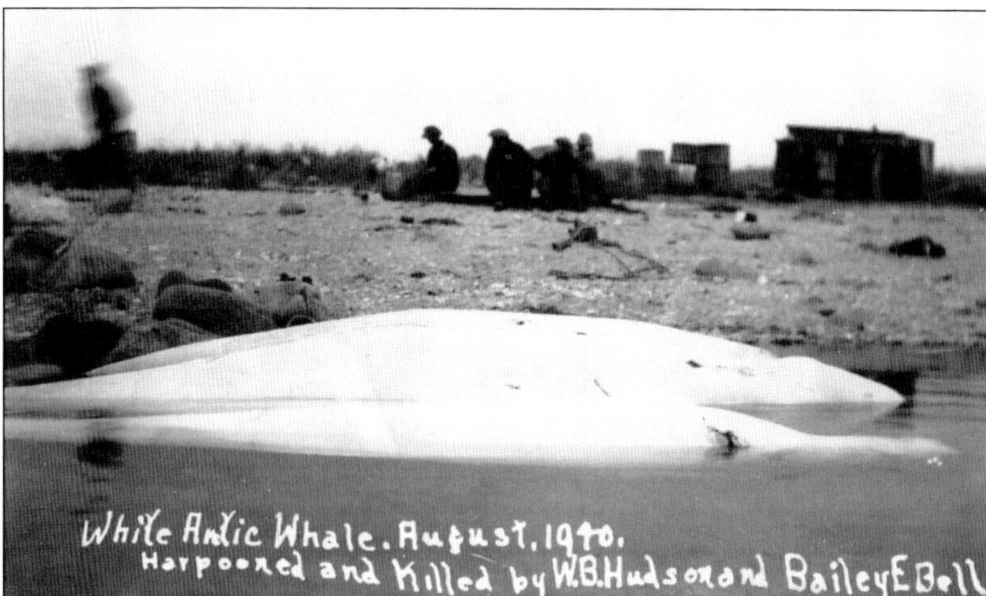

According to the writing on the photograph, this white whale was "harpooned and killed by W.B. Hudson and Bailey E. Bell" in 1940. Belugas are born dark brown or blue-gray, but their skin turns whiter as they age. By the age of nine, their skin has usually become pure white, but sometimes females retain some gray coloration up to 21 years of age. Belugas can live more than 60 years. They are medium-sized whales and weigh between 2,000 and 3,000 pounds.

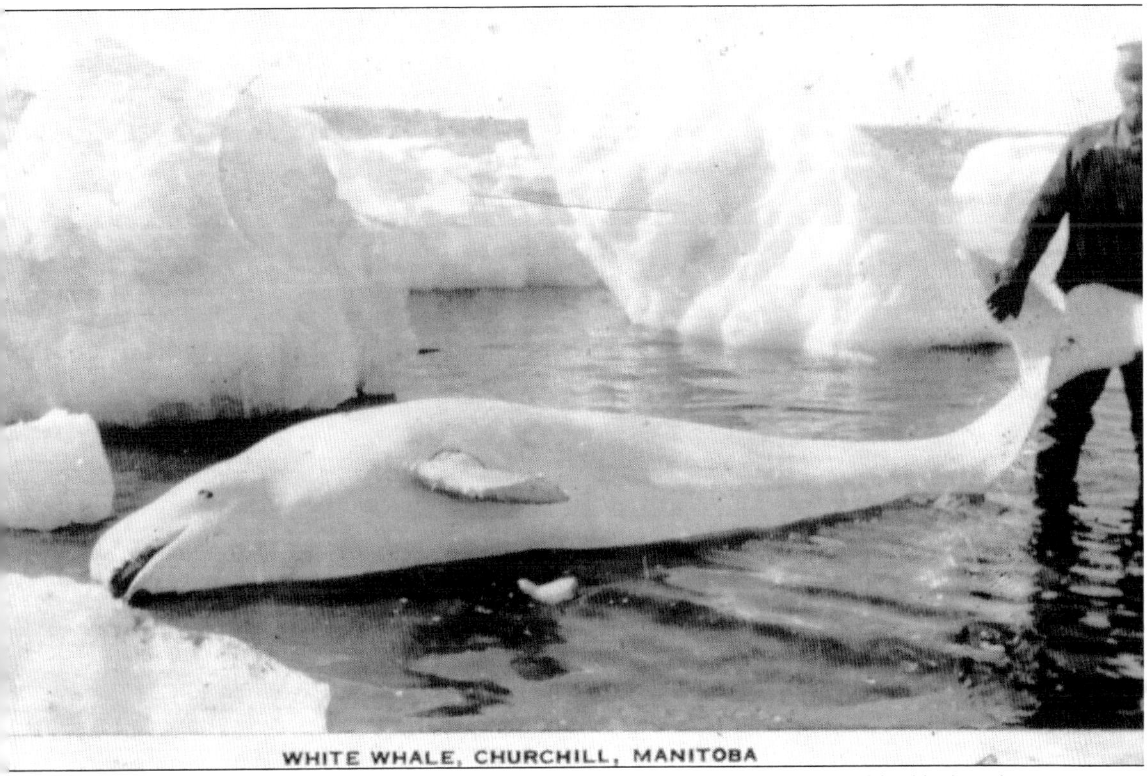

Today, beluga whales are usually taken by being shot with a rifle then quickly grabbed before they sink to the bottom. Eskimos and Indians have lived in the coastal area surrounding Cook Inlet since prehistoric times. These hunting societies use many marine resources, including beluga whales, seals, and porpoises.

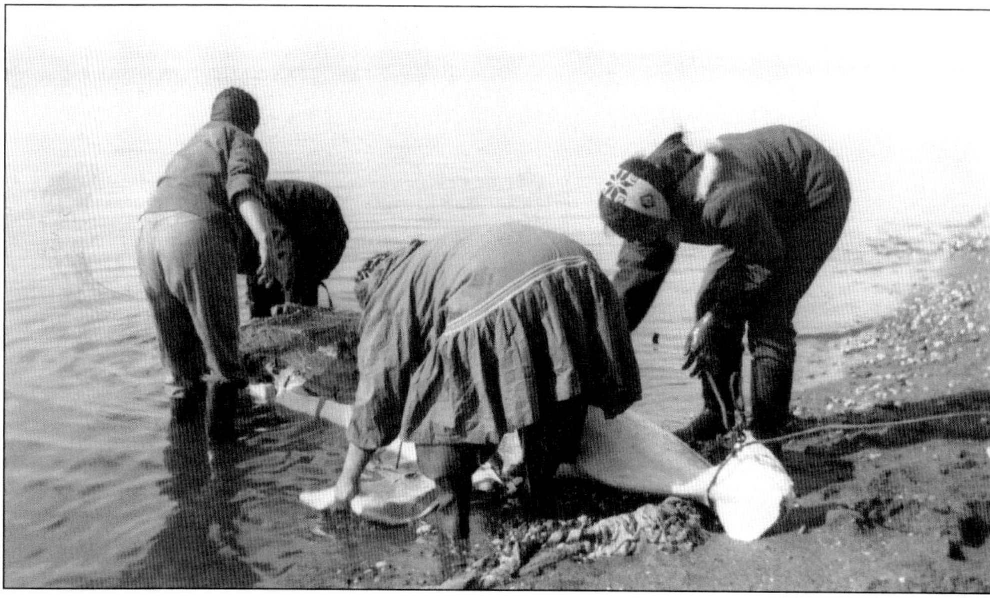

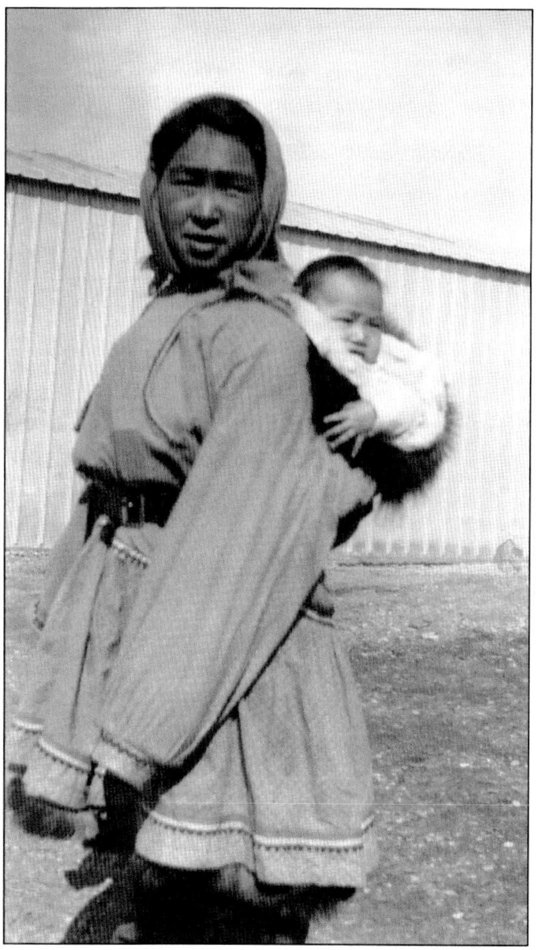

Once the men return home from the hunting, the women take over and do the work of removing the meat and blubber from the carcass. Beluga whales are very social, and they hunt and travel in close groups. They feed cooperatively around the mouths of rivers on seasonal fish.

The women clean the whales but still care for the children. Beluga whales are known to intentionally strand themselves while rubbing their skin against rocky bottoms during molting.

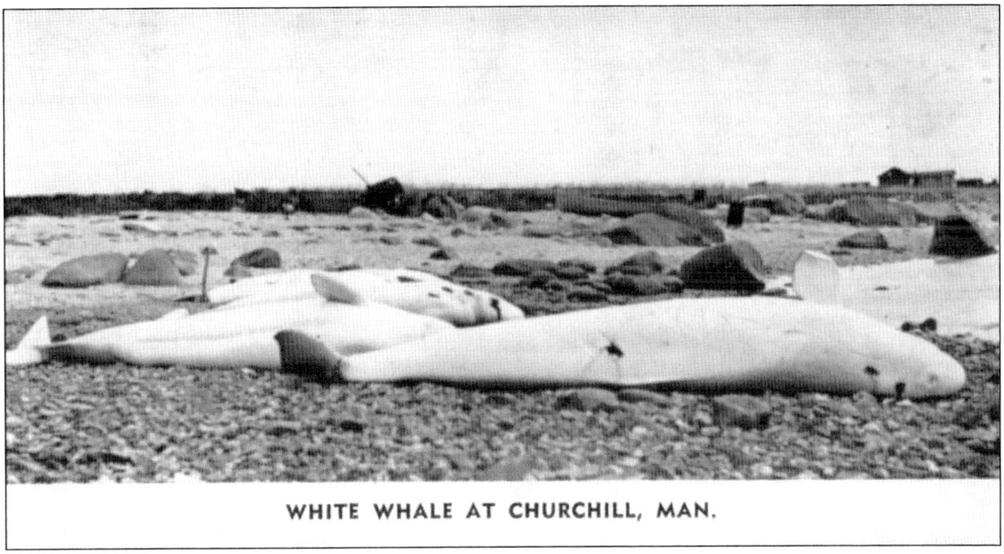

Mass stranding of beluga whales occurs in shallow beaches and river mouths, but when the tide comes in, they often refloat and swim away. Sometimes, though, stranding results in death. The belugas are known to strand themselves, either accidentally or on purpose, to avoid attacks by killer whales.

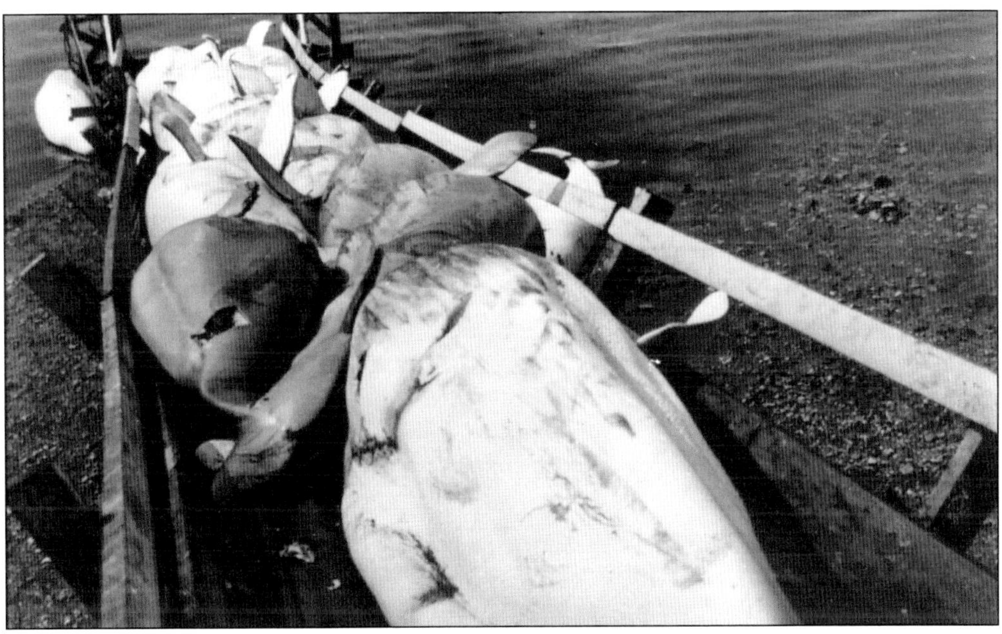

These beluga whales are lined up to be taken into the factory for processing. Natives used dams and fences to trap the animals in streams or rivers as they attempted to leave on the ebb tide. They also waited in spearing trees—driven upside-down into the mud of the river—to harpoon the whales as they swam past. The harpoons were fitted with toggle points to lodge in the whale's body and were tied to floating bladders.

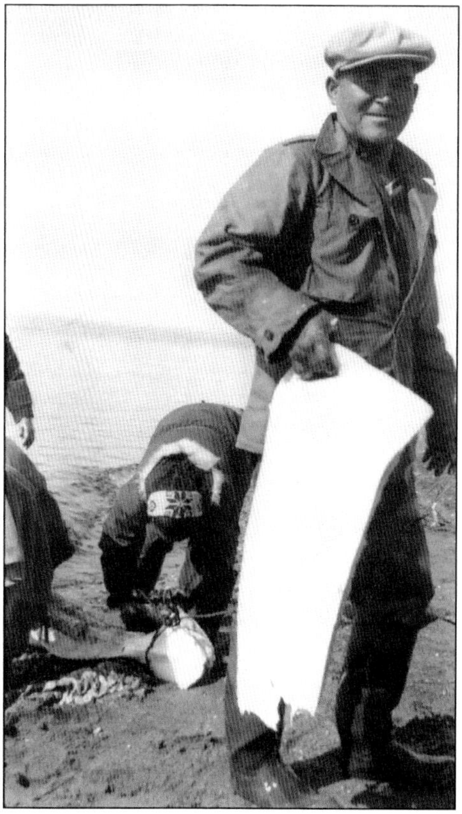

This piece of muktuk (blubber and skin) will be cut up later.

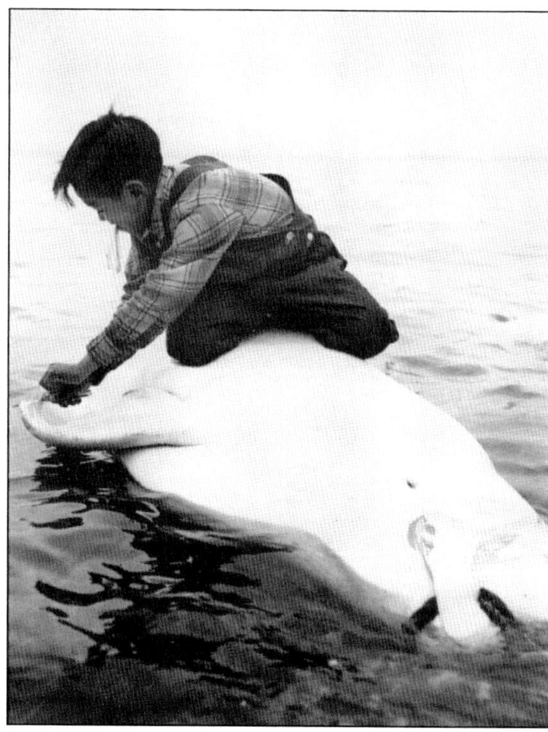

Belugas can be hunted individually from a boat or from the shore in areas where they swim along the ice in spring. This is done from Point Hope to Point Barrow. In the spring, the belugas will begin to come along the open leads about two weeks before the bowhead whales, so hunters will take some belugas while waiting for the bowheads to arrive. For a young man, shooting his first beluga is a rite of passage until he can get on a bowhead crew, then maybe one day his own boat, eventually serving as a captain of a whaling crew.

Four

Port Armstrong and the US Whaling Company

In 1910, rumor had it that in a single season a whaler caught 300 whales and produced 12,000 barrels of oil near Vancouver Island. This would have been the best season in the world except for the Antarctic. At the same time Norwegian stockholders were looking at a prospectus of a new whaling enterprise in Alaska.

Peter Bogen, one of the richest whaling entrepreneurs, raised $400,000 through stock sales to finance the Norwegian company, to be named the US Whaling Company. Two-thirds of this amount was subscribed from Scotland by shareholders in his other companies. The boats had to be registered in the United States and fly the American flag. By building the whaleboats in the United States, the company was also able to save the expense of a three-month voyage for the whale catchers from Norway.

Bogen prepared a "floating cookery," using a Norwegian steamer outfitted with the necessary machinery for handling the catch, and the three vessels needed for catching the whales. While the steamer was outfitted and shipped from Norway, the three steel-hulled chaser boats were constructed at the Seattle Construction and Dry Dock Company. "Registered in Ketchikan," writes Webb in *On the Northwest*, "as the *Star I*, the *Star II*, and the *Star III*, they soon proved to be efficient, fast boats, though like their Norwegian predecessors they were very short on human comforts."

The three whale catcher or killer boats arrived at the station at the end of April ready to begin operations. They had been preceded by the floating cookery ship *Sommerstad*, which arrived in the middle of the month. Operations were conducted both onshore and on the *Sommerstad*. This resulted in a catch during the first six weeks of operation of 71 whales. The *Sommerstad* was moved with two of the whalers to the Shumaguin Islands, where the season was going somewhat better until the US government complained about the *Sommerstad* being a foreign vessel, and it had to be moved out of US waters. At the close of the season, the *Sommerstad* sailed to Europe with the season's catch. The *Sommerstad* did not return to Alaska for the 1913 season.

The first year, 1912, the ships caught 314 whales and produced 8,500 barrels of oil—only a quarter of what had been estimated. Gross earnings were 554,000 Kroner; results for the year showed a loss of 252,000 Kroner and liabilities of 420,000 Kroner, according to Tønnessen and Johnsen.

Bogen and his directors decided to resume the hunt during 1913, and production did increase from 8,500 to 9,333 barrels of oil. Nevertheless, a loss of 85,240 Kroner had to be reported, according to Webb.

During the spring of 1913, the officers and employees had considerable problems with blood poisoning and throat trouble. Early in May, the *Star III* brought Captain Abrahamson, the manager, and four Chinese infected with blood poisoning for medical treatment. The *Star I* followed later with four more cases. Two of the men were put in the hospital. Captain Abrahamson received treatment in Petersburg.

Although the total catch for 1913 was disappointing, 75 sperm whales were secured this season by the three whalers operating out of Port Armstrong. This is the record catch of sperms for a season for any one company operating on the Pacific Coast. With sperms valued roughly at $1,000 each, the total value of the catch was $75,000. The company had found the feeding grounds of the sperms this year. Whether these were the permanent feeding grounds only future years would show.

While the number of whales caught was fairly large, the yield of oil did not come up to expectations, owing to the fact that the whales handled at the reduction plants and floating boileries failed to yield the usual proportion of oil. This year's whales did not seem to be as fat as those of previous years, apparently because of their inability to secure the proper amount of food. The Mount Katmai eruption in June was said to frighten all the whales away from that vicinity. This along with an unusually stormy season forced the whales to remain well out in deep water. The sum effect was that the whales seemed to have trouble finding sufficient food to retain their normal body fat.

Starting with the 1914 season, the three whalers and support ships with equipment and supplies would go north in May and return in September to winter quarters in Eagle Harbor near Seattle. The company would send north its whaling steamers, *Star I*, *Star II*, and *Star III*, in the early part of May. The officers and crews of these vessels worked on a percentage basis, as in previous years. The company planned to prepare oil, fertilizer, bone meal, and other products from the carcasses and would also save the whalebone.

In 1916, the station needed to be rebuilt, as it was partly destroyed by winter storms. Men and equipment went up in March for this purpose so that the station would be ready for business when the whalers and station crew arrived at the end of May. Balfour, Guthrie & Co., agents for the whaling company, sent Capt. E. Earling, harpooner of the steam whaler *Star III*, to Port Armstrong to make a survey of the damage preliminary to the adoption of plans for the reconstruction of the station.

The whaler *Star I* was ordered delivered to the Navy on December 24, 1917, and it was acquired on January 19, 1918, commissioned as USS *Star I*, and assigned ID No. 1351 on March 30, 1918. After service as a minesweeper in the Pacific Northwest during and shortly after World War I, the vessel was returned to the US Whaling Company on February 12, 1919. Sister ships *Star II* and *Star III* were not taken over by the Navy. In 1918, all three whalers were back at work. The company took 121 whales, far off the 300 average of some of the other stations.

The whaling companies were devastated in 1920 by woefully low whale stocks, exorbitant operating costs for the previous six years, and an end to the high wartime demands. Profitable catches would not resume on a regular basis until the mid-1920s.

In 1922, the company took 117 whales, including 60 sperm whales, and only produced somewhat over 6,000 barrels of oil. The 1922 season was the last for the US Whaling Company. It was reported early in the year that Buchan & Heinen, who had leased the whaling station at Port Armstrong, had ordered a new reduction unit from the California Press Manufacturing Company and expected to engage in the production of herring oil and meal.

In October 1923, it was reported that *Star I*, *Star II*, and *Star III* and 36 Norwegian whale men under the command of Capt. Gustav Petterson sailed from Seattle. The three small steamers were sold to a Norwegian whaling company. The vessels sailed for Auckland, New Zealand, which was to be their base of operations for work in the Antarctic.

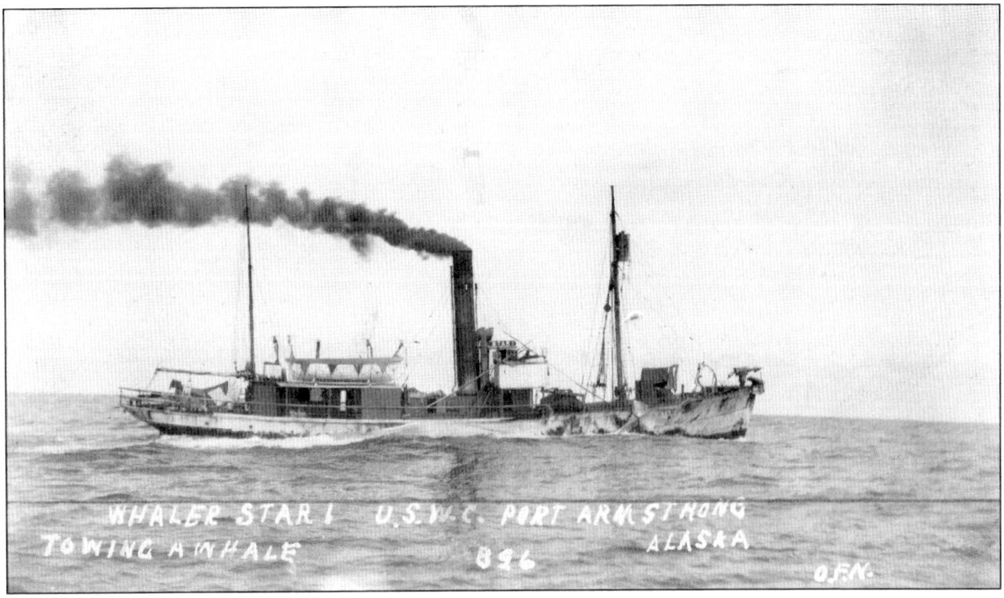

Above, the *Star 1* tows a harpooned whale to the Port Armstrong station. (Courtesy of University of Washington Libraries, UW 33489.)

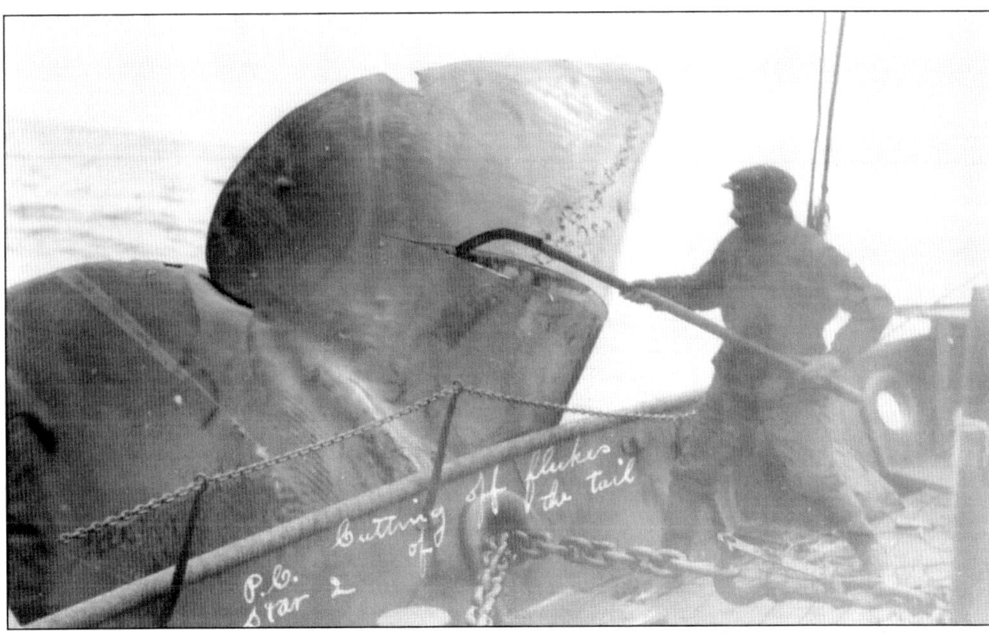

This image shows a worker on the *Star II* as he cuts the tail off a whale. (Courtesy of University of Washington Libraries, Special Collections, UW 27653.)

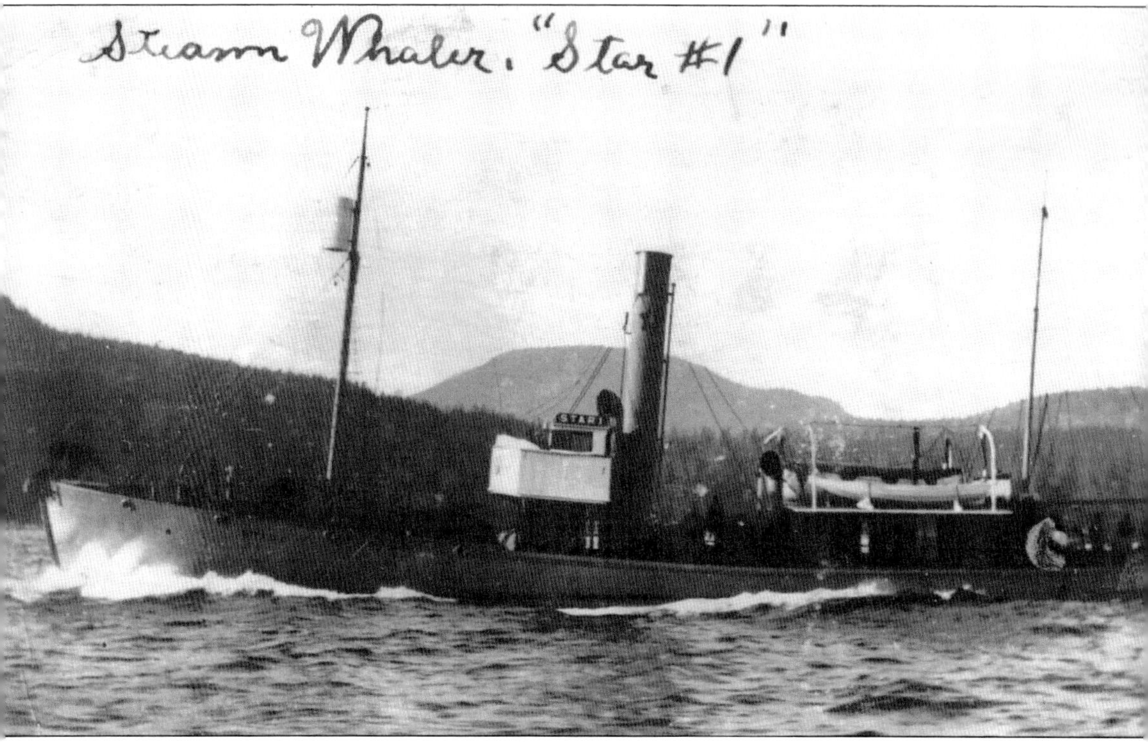

Star I was a 196-ton steam-powered whaler built in Seattle in 1912 for the US Whaling Company. It was chartered by the Navy in 1917 and placed in commission in late March 1918 as USS *Star I*. After service as a minesweeper in the Pacific Northwest during and shortly after World War I, It was returned to its owners in February 1919.

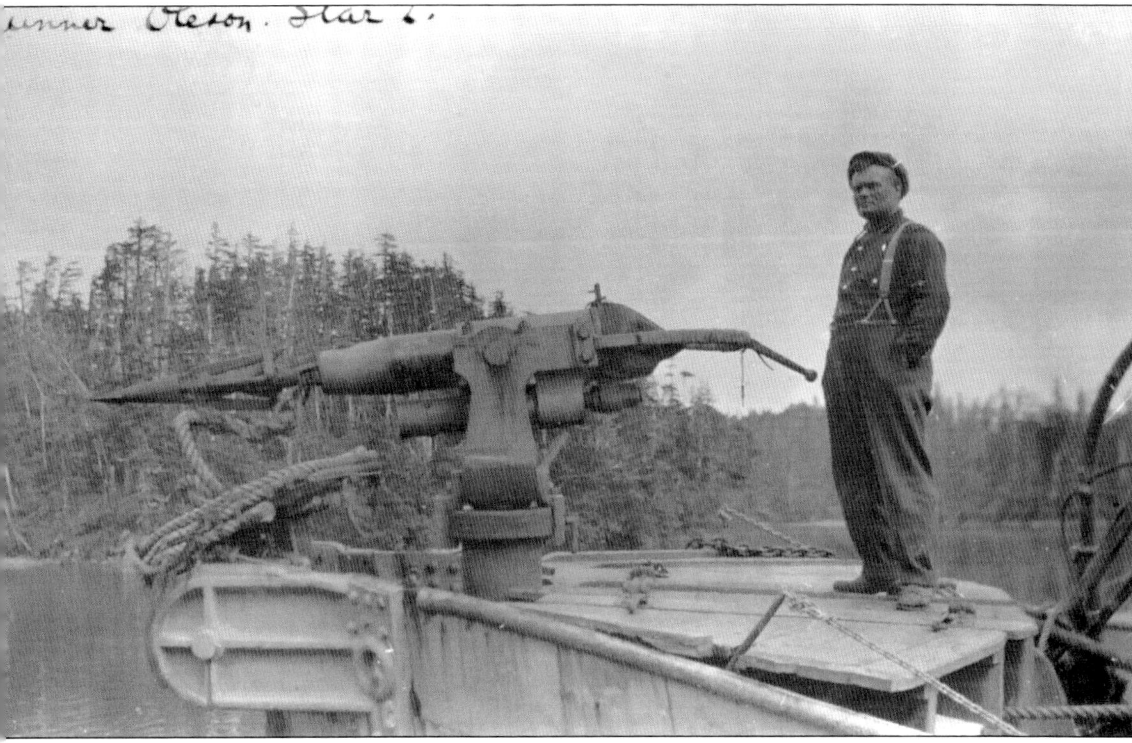

Star II and *Star III* were identical but slightly smaller than *Star I*. The two boats were 95 feet long and weighed 71 tons, while *Star I* was 117 feet long. A Mr. Olson was the gunner for a while on *Star I* and possibly a few years later on the whaler *Tanginak*. In October 1923, it was reported that *Star I*, *Star II*, and *Star III* were sold. The captains and crews—36 men total—left Seattle and went to New Zealand to work in the Antarctic. The vessels names were probably changed, as no further information can be obtained as to their further history.

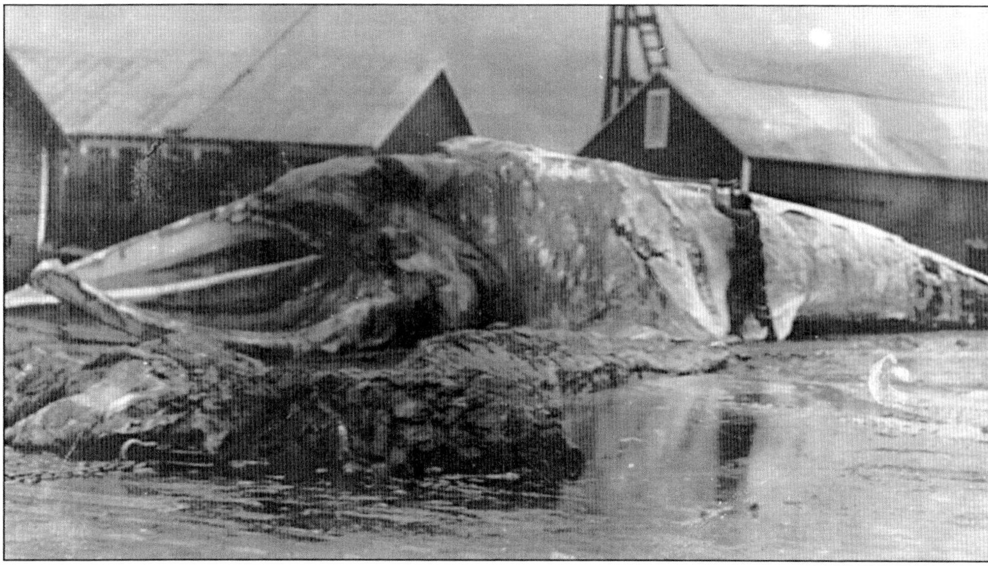

This finback whale is being flensed on the platform at Port Armstrong. The blubber has already been removed in strips and then cut up and taken to the boiler to "try out" as much of the oil as possible. Blubber is much like bacon fat on a large scale; when it comes up to temperature, the oil comes out as a liquid. As one continues to put in more blubber, it stays hot and, with all the boiling liquid, continues to cook down.

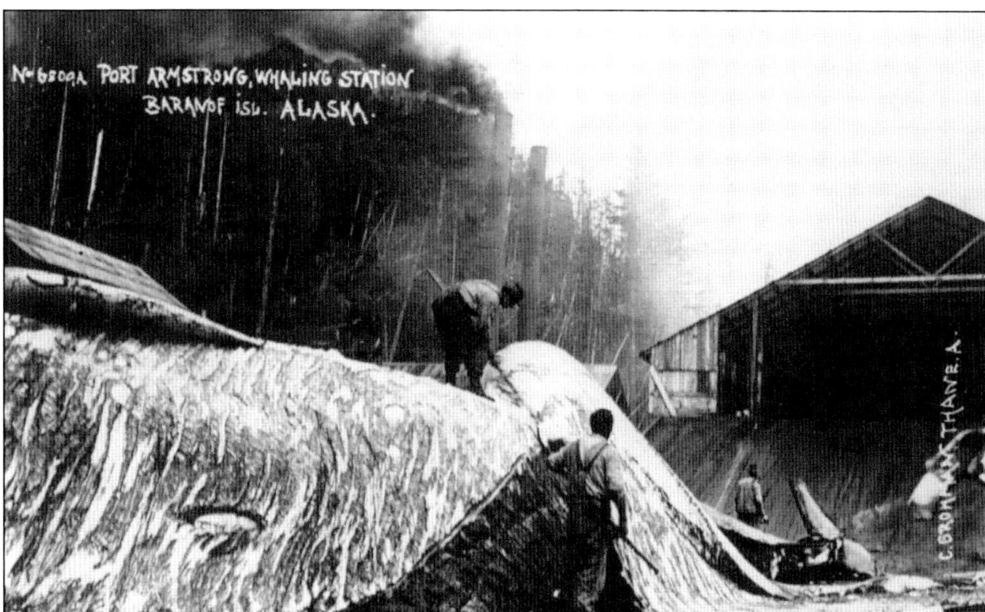

This image shows a sperm whale being flensed. Sperm whale fat looks different than that of baleen whales; it does not stay in the form it is in and therefore cannot be cut into blocks. One reason this may be so is that there is far more oil in the meat and fat than in the bodies of other whales. Sperm whales are not used for food for humans or animals.

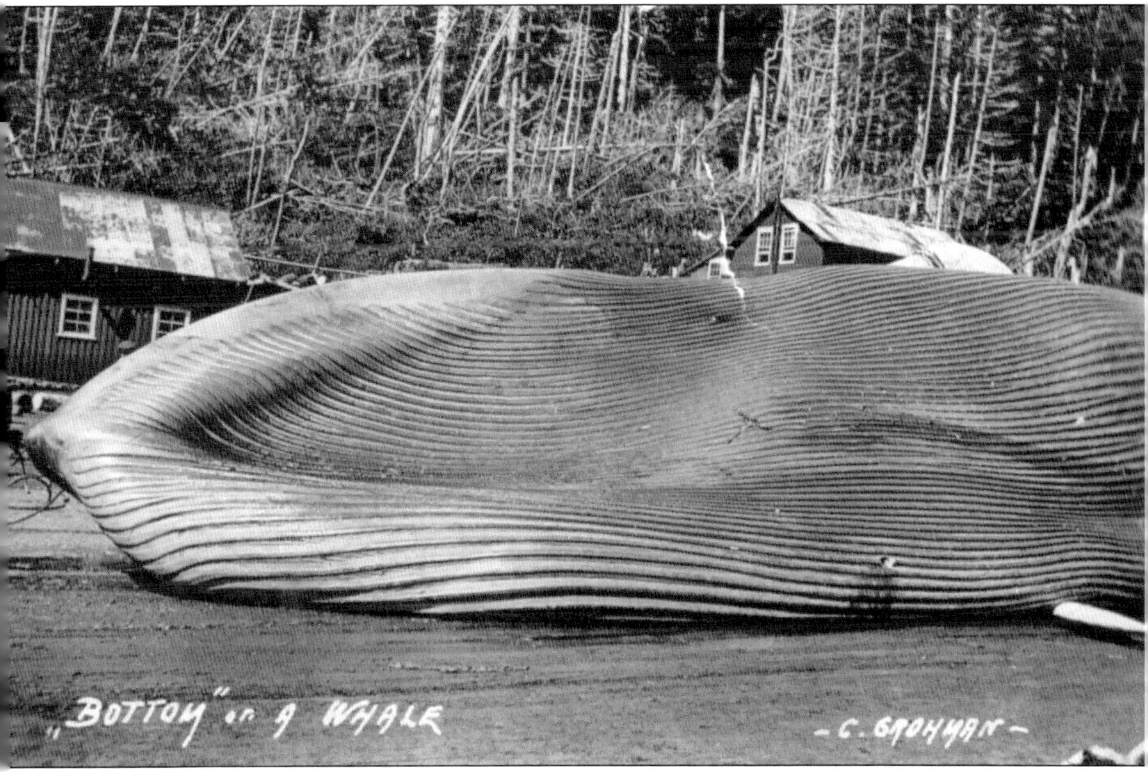

This whale's fins are not visible, making it hard to identify the species. Grooves in the lower throat area act like pleats in a dress to aid expansion. Baleen whales open their mouths and take in a very large mouthful of water loaded with nutrition, which could be krill, fish, worms, or whatever is available.

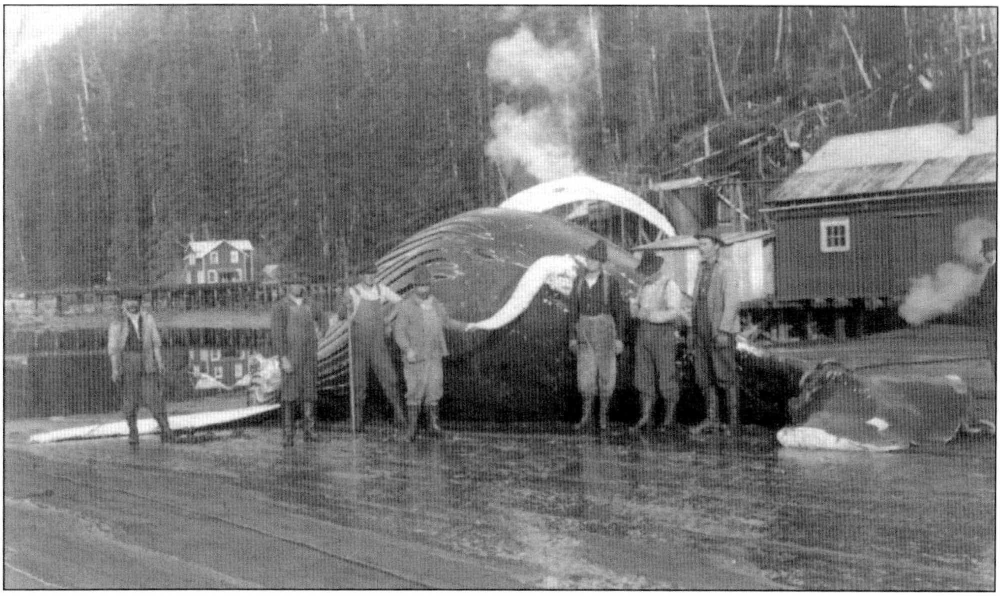

Men pose here with a male humpback whale. The man just left of center is holding the tip of the whale's penis.

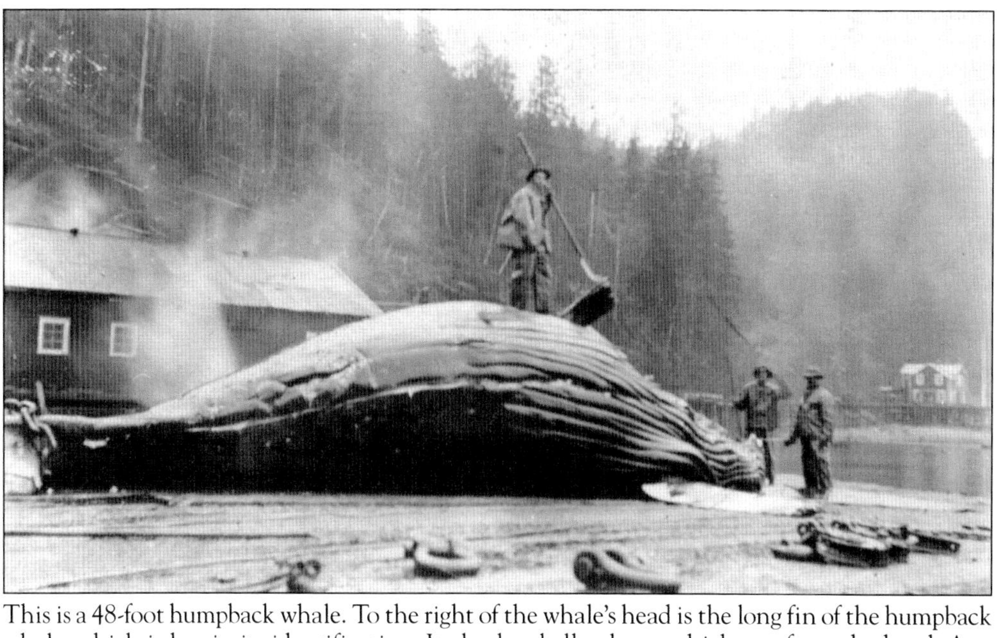

This is a 48-foot humpback whale. To the right of the whale's head is the long fin of the humpback whale, which is key in its identification. It also has belly pleats, which run from the head. Any whale with these pleats will have baleen in place of teeth, indicating that it strains food.

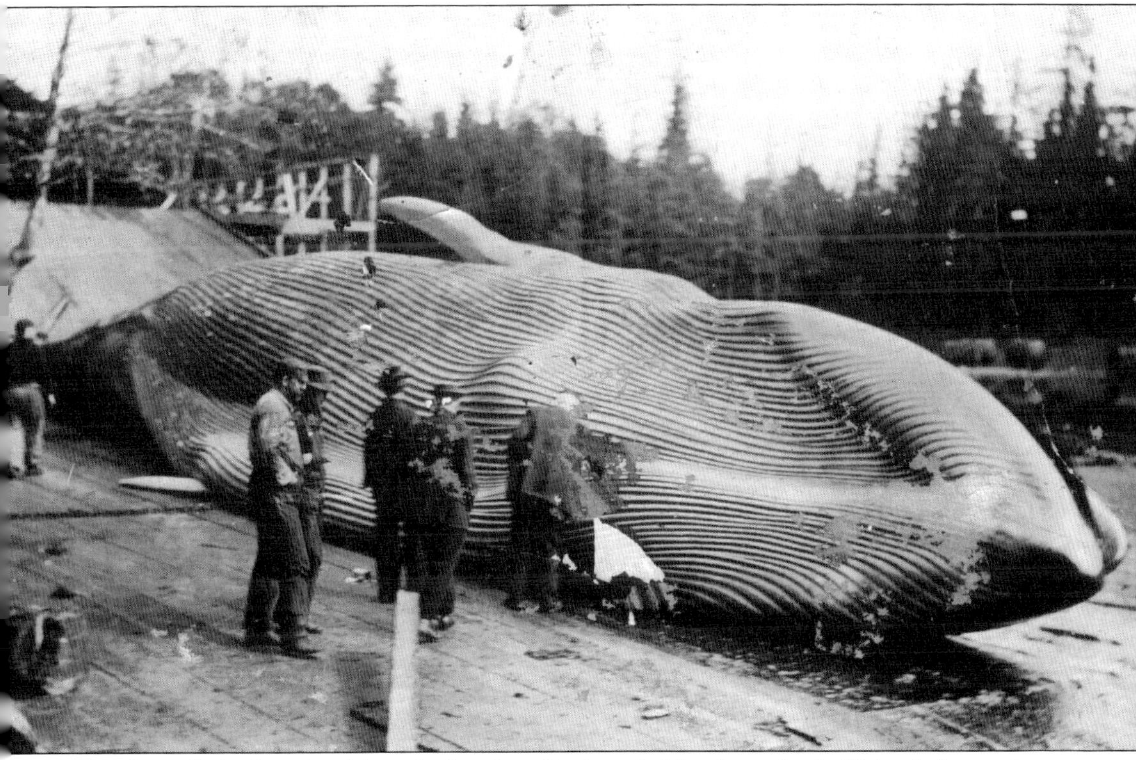

It is hard to tell the length of the side fins, but this is probably a humpback. This whale may be close to 60 feet in length, so at one ton per foot, one could estimate the weight at 60 tons. Large whales like finback whales may reach 70 feet, and blue whales may reach 90 feet, with weight estimated at 1.5 tons per foot. Smaller whales—40 feet or less—are estimated at one half-ton per foot.

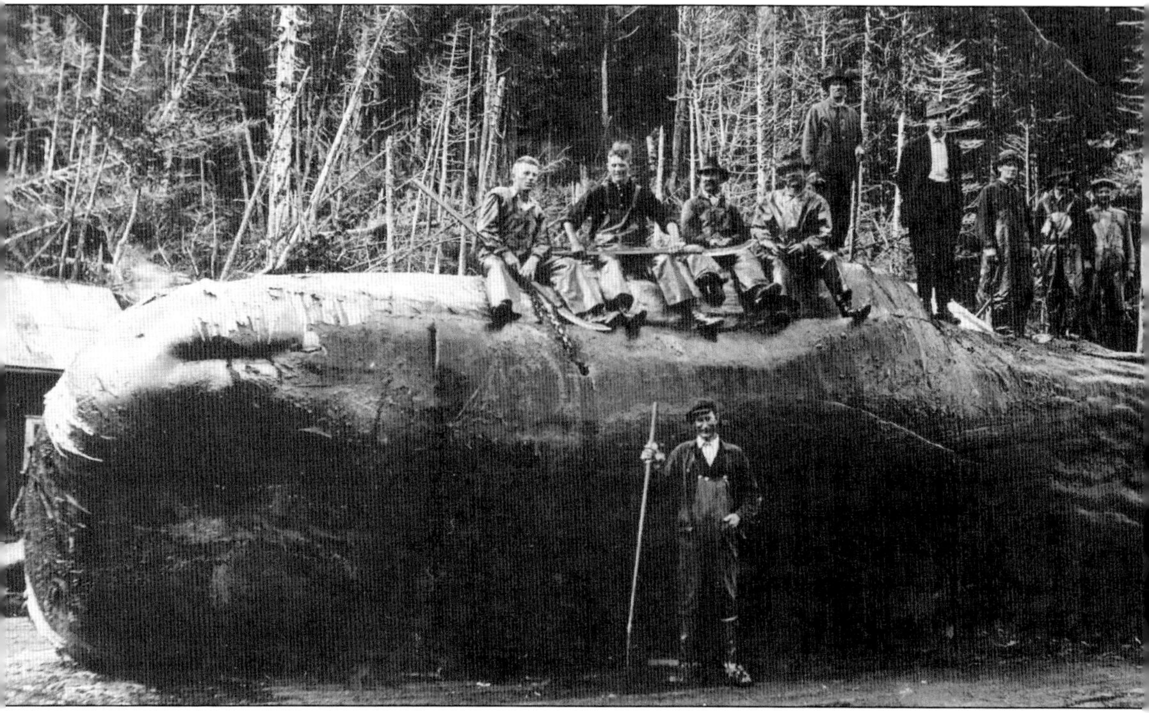

This sperm whale head is to have the oil removed. The sperm whale's head contains a sack that holds the oil, with large blood vessels all around the organ. It is said to help the whale in echolocation, which is a type of sonar. Scientists have also theorized that the whale can use the oil when diving. It has been said that the whale can channel blood flow away from the head, which would cause the oil in the head to solidify and act like a diver's weight belt, making the whale negatively buoyant at a shallower depth.

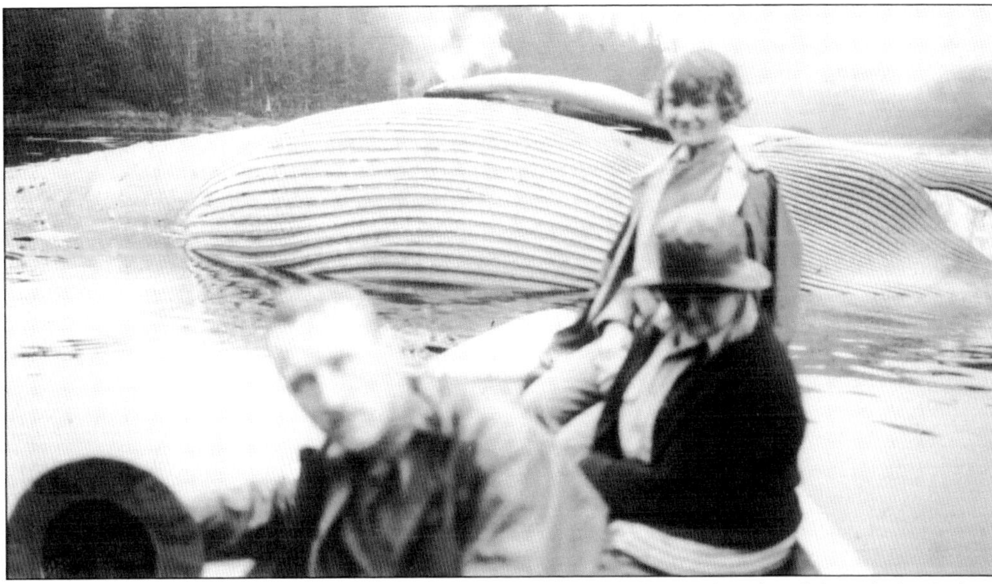

This photograph shows a baleen whale; note the pleats on the underside of the whale. Baleen whales taken would be humpback and finback whales. Compare the woman's size to that of the whale—this one is probably around 60 feet long. The trees in the background help to identify Port Armstrong, as Akutan is barren like the surface of the moon.

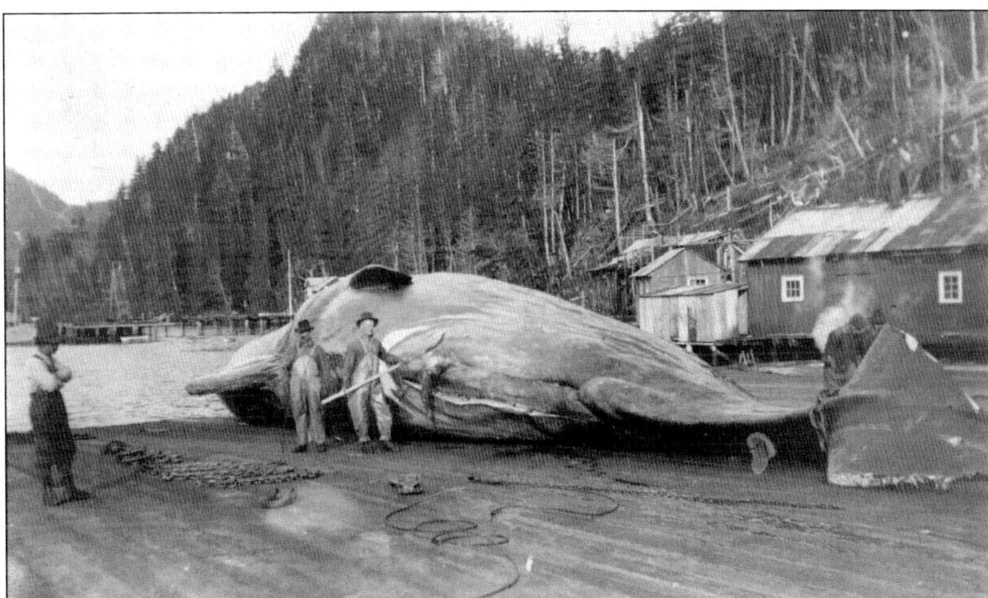

Here is another sperm whale, which was more common—along with blue whales—at Port Armstrong than other whales. Farther west, cities like Akutan received many more humpback and finback whales. Although Port Armstrong did not get as many whales, the ones that did appear in the area were larger. Note the jaw of the sperm whale and the head flenser with his blade showing the whale's penis. The building with the white windows helps to identify the station.

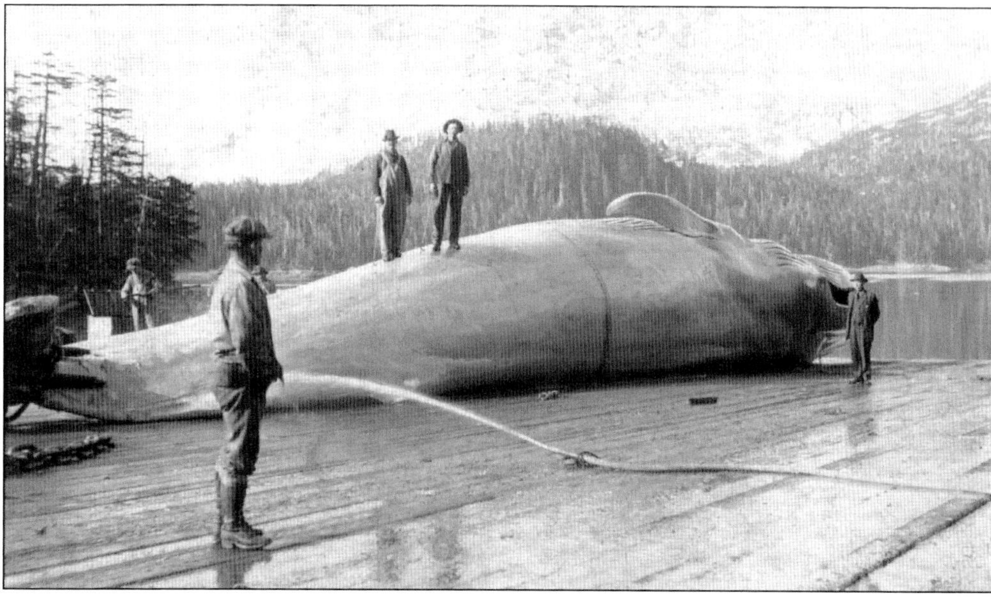

A baleen whale just brought in on the ramp gets a bath to remove particles of dirt and rocks accumulated while it was dragged ashore. Rocks will dull the blades of the knives used in flensing. The hills and trees are indicative of the area around Port Armstrong.

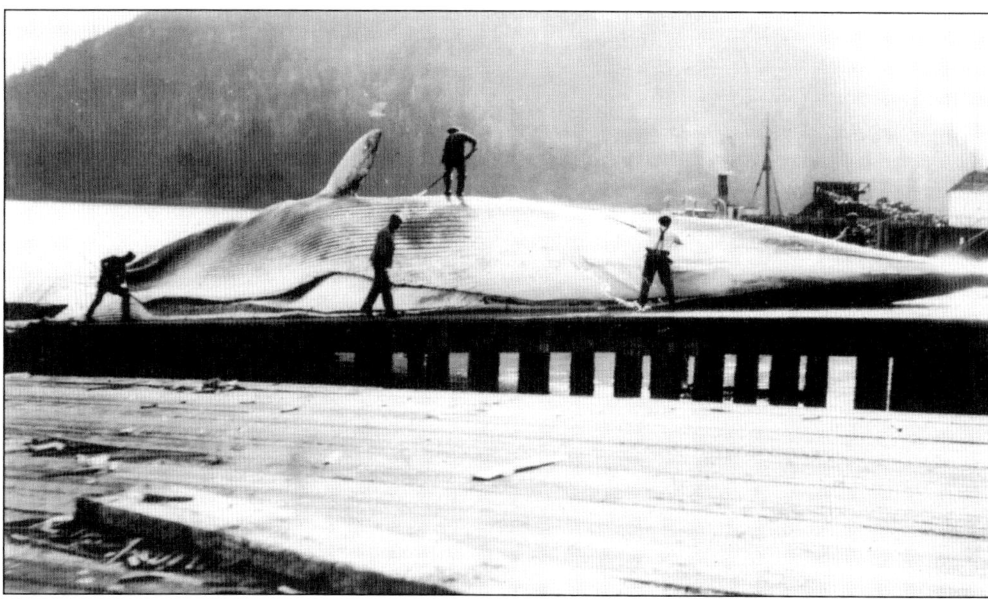

Here, the flensing ramp at Port Armstrong has a blue whale on it. The blubber will be pulled off in long strips at the station. The flensers make long cuts, and then a cable tied to a winch is inserted via hook into the blubber near the head of the whale. As the winch is started, the flenser will separate the blubber from the meat and pull the blubber off from head to tail.

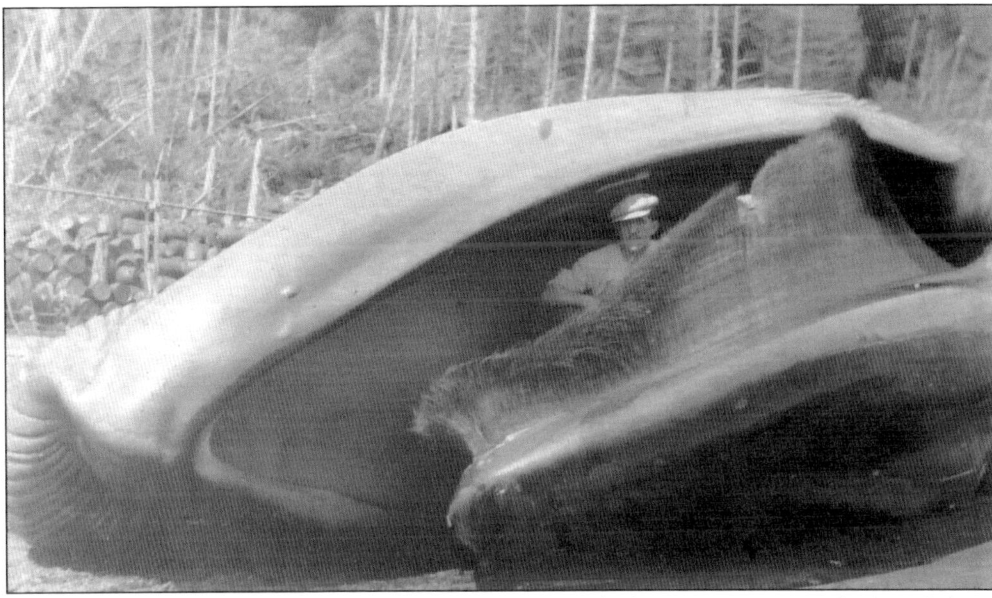

It was not unusual for tourists to find a way to the station by boat, since it is not far from Sitka or Juneau. Navy people in the area and scientists also came to Port Armstrong during the season. A favorite photograph seemed to feature someone in the mouth of the whale, showing the baleen plates. The baleen itself is clean, not oily or greasy, but it could still have an odor.

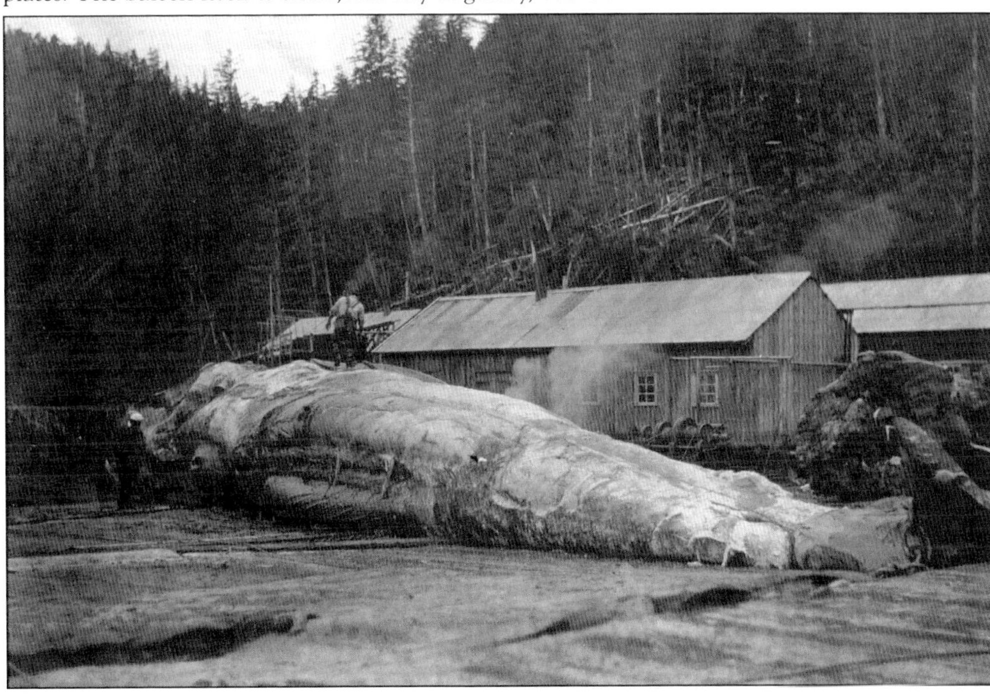

This whale has had the skin and baleen removed. The next operation is to remove the meat and then dismember the bones. The meat and bones are processed into fertilizer. There was and still is not much use for the meat. It has been used for dog food or mink and fox feed, but there was so much available that it was used mostly to make fertilizer. Some nations (Japan and Norway) still hunt whales and use the meat for food.

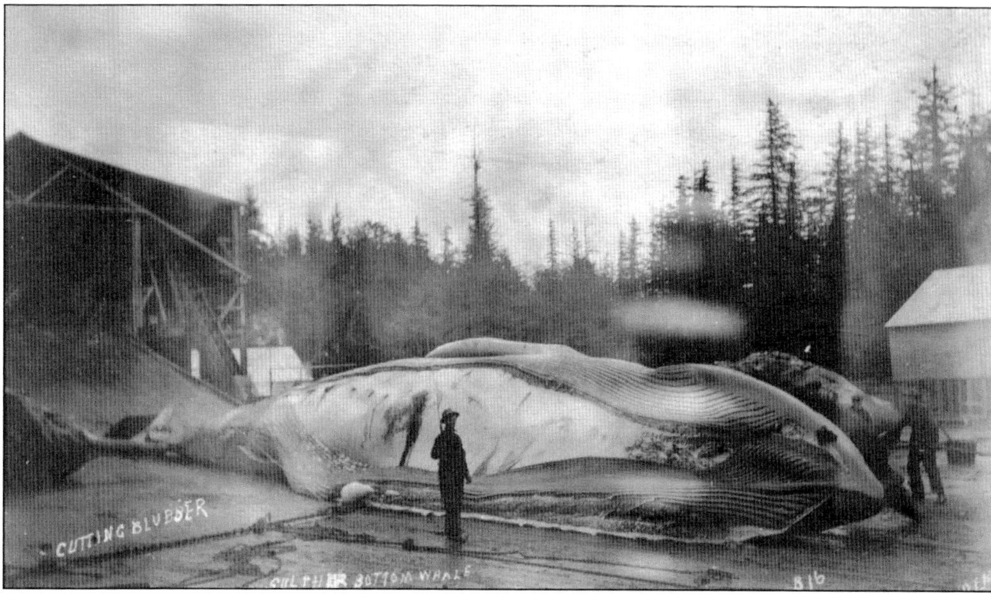

A "sulphur-bottom" (blue) whale is being flensed to remove skin and baleen. The lines near the head on baleen whales are actually pleats that expand to allow the whale to ingest a large capacity of water and fish, then close its mouth and force the water and fish through the baleen. The water will go through, but krill and fish will be caught in the baleen, and once the water is gone, the whale uses its tongue to wipe off the baleen, then swallows the food.

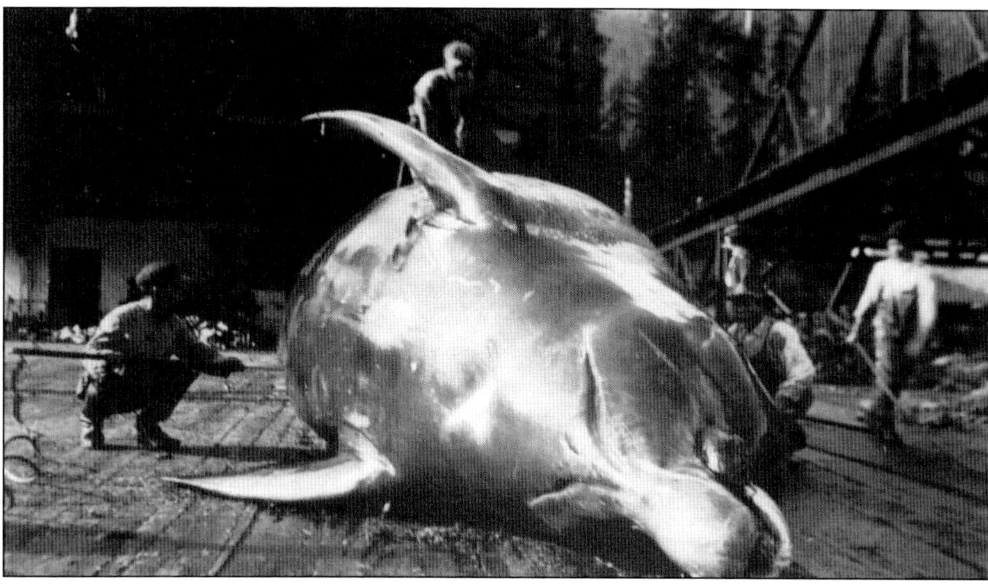

This whale, identified as a North Pacific whale, was probably a bottlenose whale, which is rarely taken by whalers. It may have been taken for scientists, which was not unusual in the 1910s to 1920s. Scientists were able to visit whaling stations to see and research fresh specimens. Before this became routine, a scientist would have to stay with the whale ship for a whole season unless the ship made port, which was not common.

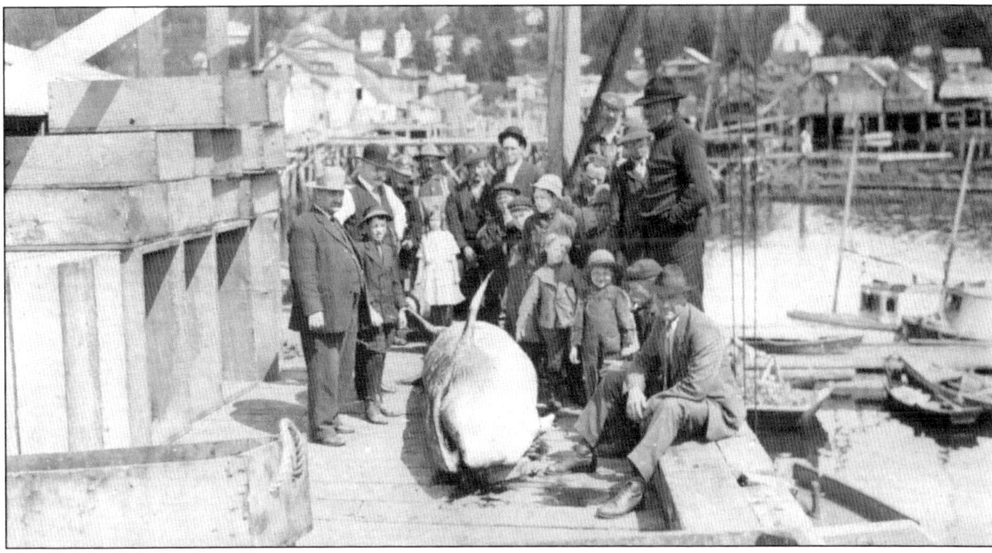

This was a finback whale fetus. It was not too common for a whale to have a fetus, but when it was available, it would be utilized much like the larger whales. Sometimes the fetuses were given to universities or biologists for study.

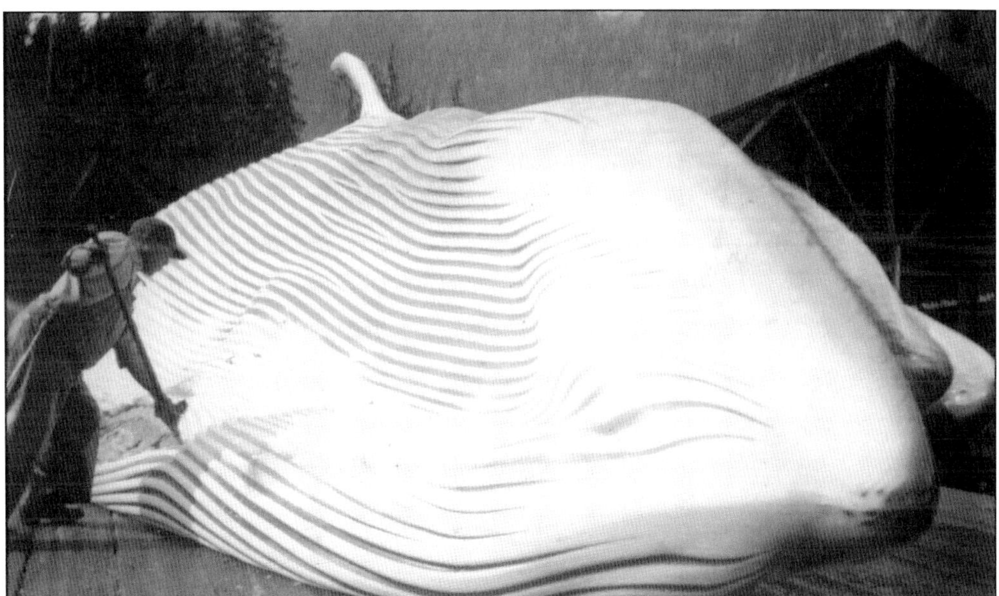

A flenser is making a first cut on this baleen whale. The pleats are better seen from this viewpoint. The blubber can vary in thickness based on the health of the whale. If the whale was sick and not eating well, it could have very little blubber. In some seasons, nature interfered with the whaling. In years when there was volcanic activity, whales were known to be thin. One thought was that the volcanic activity might kill off the krill.

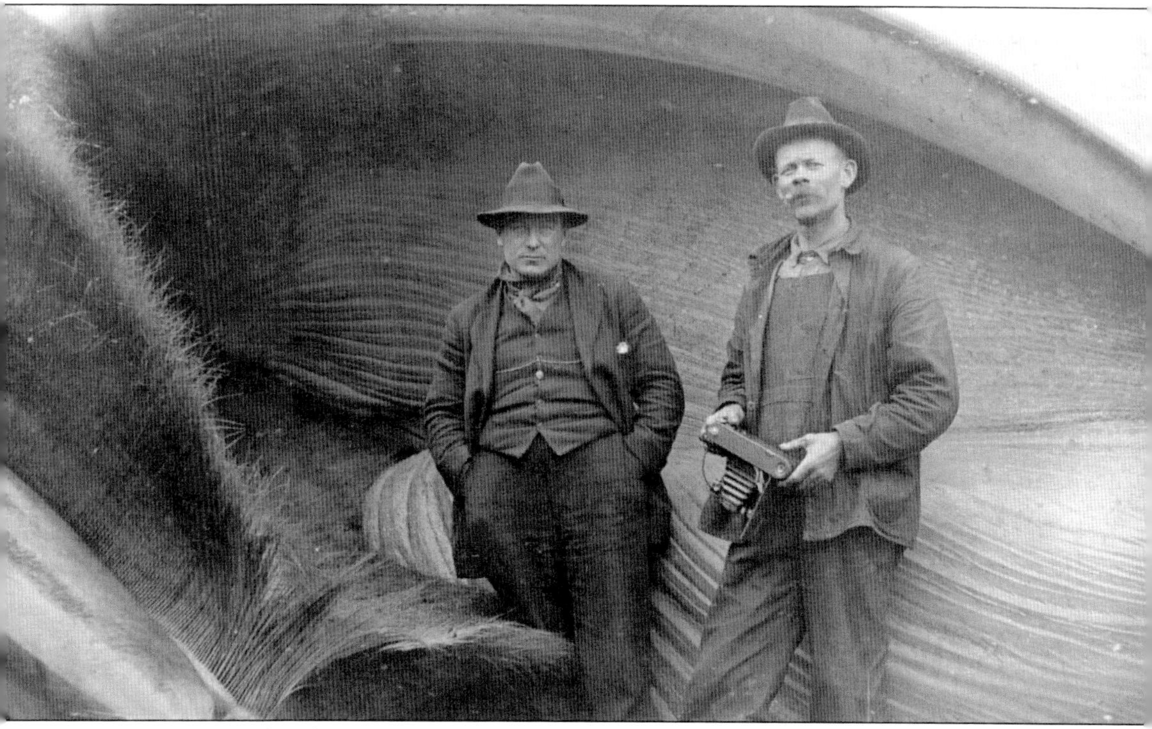

Two men stand in the mouth of a whale. While these may have been tourists, the man with the camera could be one of a number of photographers who visited the station, or he could have been from one of the national magazines, which often sent correspondents to report on whaling.

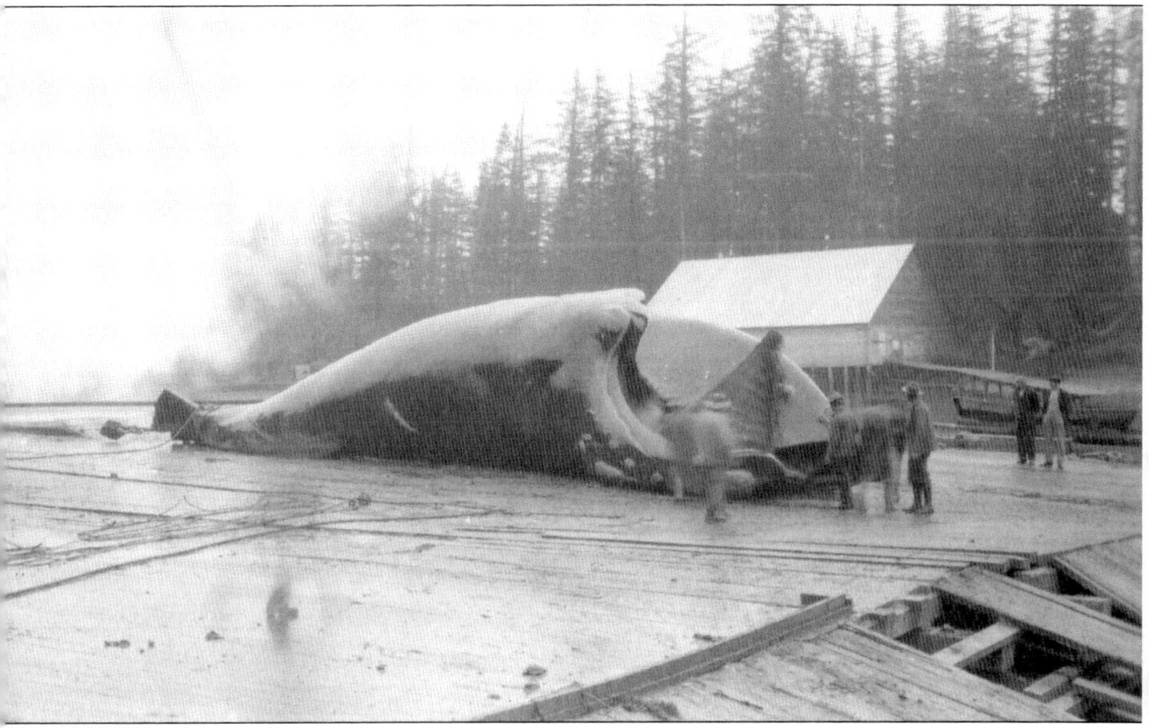

Here is another 1915 photograph of men flensing a whale at the Port Armstrong whaling station. (Courtesy of University of Washington Libraries, Special Collections, UW 33492.)

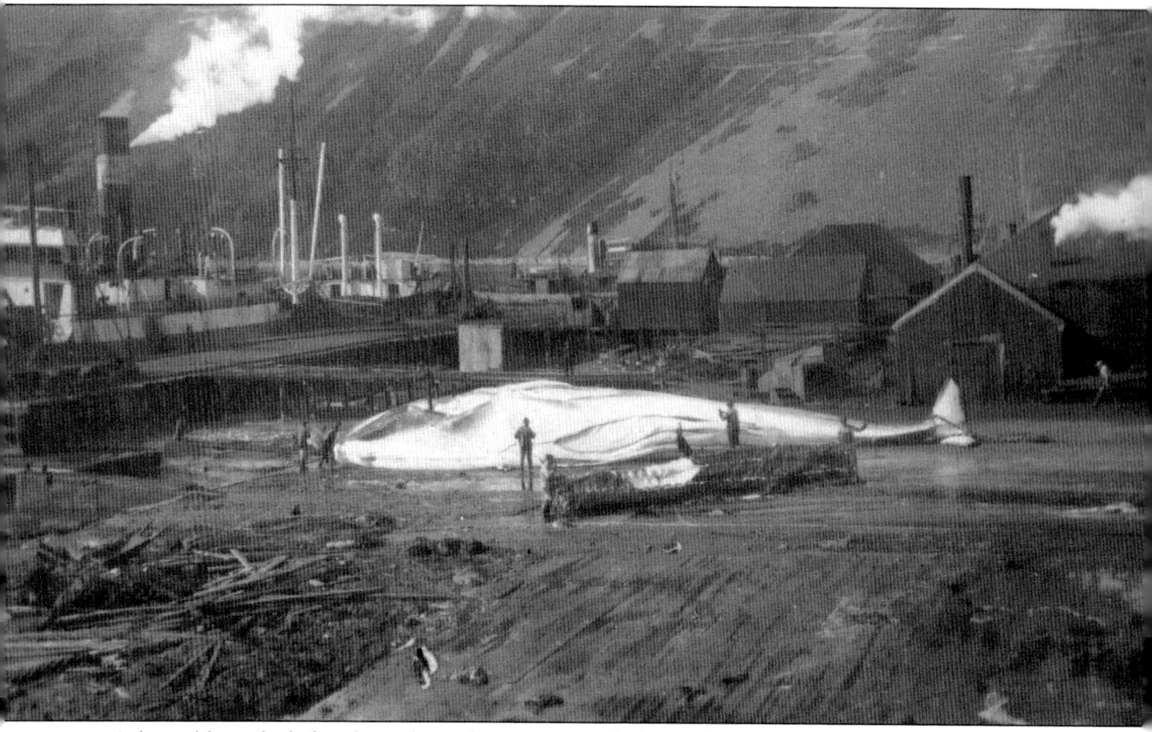

A large blue whale has been brought in to an Alaskan whaling station and has been hauled up the ramp in order to start the flensing process.

Five

Akutan Whaling Station, Part One

The Alaska Whaling Company was established by Lars Christensen of Sandefjord, Norway, in October 1911 with a capital of $315,000. A third of the capital stock was raised in Norway, and the rest was subscribed in the United States from Norwegian Americans. Christensen intended to operate a floating factory in addition to a shore station.

Moran's Shipyard in Seattle won the contract to build two whaling steamers to be named the *Unimak* and the *Kodiak*. The steamers were to be 100 feet long and weigh 100 tons. The whaling station was to be built on Akutan Island due to its proximity to the whale migration routes. Akutan Island itself was fairly inhospitable, with few trees and other vegetation. The island was mostly volcanic rock. According to Webb, the whaling station was to be built near the docks, with the living quarters and other support buildings spread out on the rising land behind the station.

The supply barge *Coronado* left Seattle for Akutan on August 15 in tow of the tug *Goliah*. The vessel had a cargo of coal, crude oil, and general merchandise for the company's station. The *Coronado* was to remain in the north until fall, when it would return south under its own sail with a cargo of whale oil.

April and the early part of May brought the launching at Seattle of many of the new whaling steamers going into commission in the North Pacific. The whalers *Unimak* and *Kodiak* were launched by Moran's.

The Alaska Whaling Company had a series of problems with its supply ships, starting in May 1912, when the tug *Goliah*, with the bark *America* in tow loaded with construction materials, whaling equipment, and supplies, was en route to the station. Upon arrival at Akutan, it reported the loss of it mate, who fell overboard while the vessels were about 450 miles off Cape Flattery.

The old bark *Haydn Brown*, reduced to a barge, was lost while returning to Puget Sound after delivering supplies for the Alaska Whaling Company. The *Brown*, in charge of Capt. L.D. Peterson with a crew of seven, was in tow of the tug *Pioneer* under Capt. Thomas Nielsen when a tremendous gale was encountered off Cape St. Elias, Alaska. Only one of the crew survived. H.W. McCurdy's 1966 *Marine History of the Pacific Northwest* recounts the struggle:

> After battling the storm for 48 hours, the *Pioneer* was in such danger of foundering that it was necessary to cut the tow loose. On the night of May 12, the barge piled up on the rocks off Montague Island, and the crew was swept overboard by the mountainous seas. August Palmer and another seaman reached shore, but only Palmer survived.

Besides Captain Peterson, those lost were Karl Westerlund, donkeyman; Chris Jansen and J. Martin, seamen; K. Kawahara, cook; and two unidentified passengers who had been taken aboard at the whaling station.

The *Admiralen* was one of the first three floating factories fitted with radio telegraphy. According to Tønnessen and Johnsen, "Its engine was so underpowered that it took as much as four months to complete the voyage from Sandefjord to Akutan." In the *Honolulu Star-Bulletin* of July 9, 1912, an article stated that "Reports from the north are to the effect that the innovation of a 'whaling station on the water' which was made by Norwegians off the Alaskan coast has been a success." The station was made of the steamer *Admiralen*, which, on its way to Akutan some weeks prior, created considerable excitement while in a gale off Humboldt by sending out many SOS signals after it lost its rudder in the heavy seas.

The *Admiralen* was then towed to San Francisco and repaired, having been battered by the storm. The vessel was fitted with refining tanks and averaged 200 barrels of oil a day. The *Admiralen*, a steamer of 1,517 tons displacement, was remodeled at considerable cost to the Norwegian owners, but reports stated that gratifying revenue would be returned from the investment during the whaling season. The *Admiralen* was among the first vessels to attempt the unusual operations attendant upon securing the oil on board instead of on shore. Other companies were said to be thinking of doing the same the next year, believing that considerable savings could be effected through the new scheme. The *Admiralen* was repaired and reached the Unalaska station from San Francisco. Reduction on board *Admiralen* started on June 3, followed by the shore station on August 24. Both stations wrapped up operations on October 21.

The old theory that troubles never come singly seems to have been borne out in the case of the Alaska Whaling Company, for following the temporary disablement of the company's floating factory *Admiralen* and an accident in launching its whaler *Kodiak*, the barge *Haydn Brown*, carrying the supplies for the northern station, was wrecked in Alaskan waters, seven of her crew of eight men being drowned.

Akutan was not a good location for living and created problems for the employees. It can be seen from the Akutan photographs that the location was barren of trees and was almost all rocks. The location was far from shipping and supply lines, which created higher shipping and resupply costs. The first Akutan plant was run by Norwegian managers who were far from the action, which caused delays in receiving instructions for the operations. The catch returns in 1912 were not up to expectations, but the reason can be placed on the Katmai eruption in 1912 in southern Alaska. The volcanic activity had the result of killing most of the krill, the baleen whales' basic food.

Whaling was curtailed at the Alaska Whaling Company at Akutan after the 1912 season. Bob Webb, in *On the Northwest*, attributes the failures of the station to be the station manager, who was hired locally, as well as "time, distance and changeable weather." The station manager was a Norwegian American pastor and had no knowledge of whaling operations. Christensen lost around $150,000 in the 1912 season, and there was not a 1913 season. The whaling company sold the *Coronado* to the Canadian Pacific Railway Company, which lost it near Sechelt, British Columbia, about seven months later.

In 1912, the Alaska Whaling Company was reorganized as the North Pacific Sea Products Company, and the company's board was entirely American. However, even though the catches were good, especially from 1915 to 1918, the reconstituted company could not balance its books. According to *The History of Modern Whaling*,

> In 1917, sales of oil netted $600,000 at £41 pounds a ton, while guano accounted for $46,410; expenses totaled $225,400. Shares soared, and good dividends were paid. The general meeting of January 1918 was pervaded by an atmosphere of euphoria. . . . The managing director was visited by a gentleman who expressed an interest in purchasing the company. The American shareholders, who held a majority, agreed to the sale, whereas not all the Norwegians were of a like mind. They were to regret this, as did the purchaser, the Victoria Whaling Co., which had earlier absorbed the American Pacific Whaling Company.

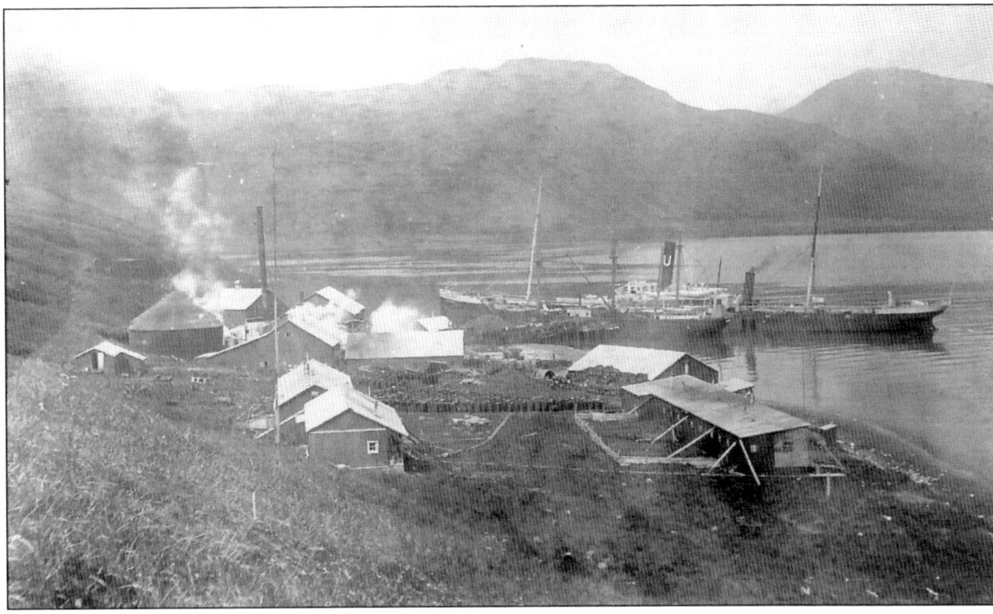

This is an early view of the Akutan whaling station, probably taken during the season or at the start of the season. The bark *Fresno* is in position at the end of the dock. Note the steamer on the outside of the *Fresno*; it could be there to pick up or drop off employees and take back any product that is available.

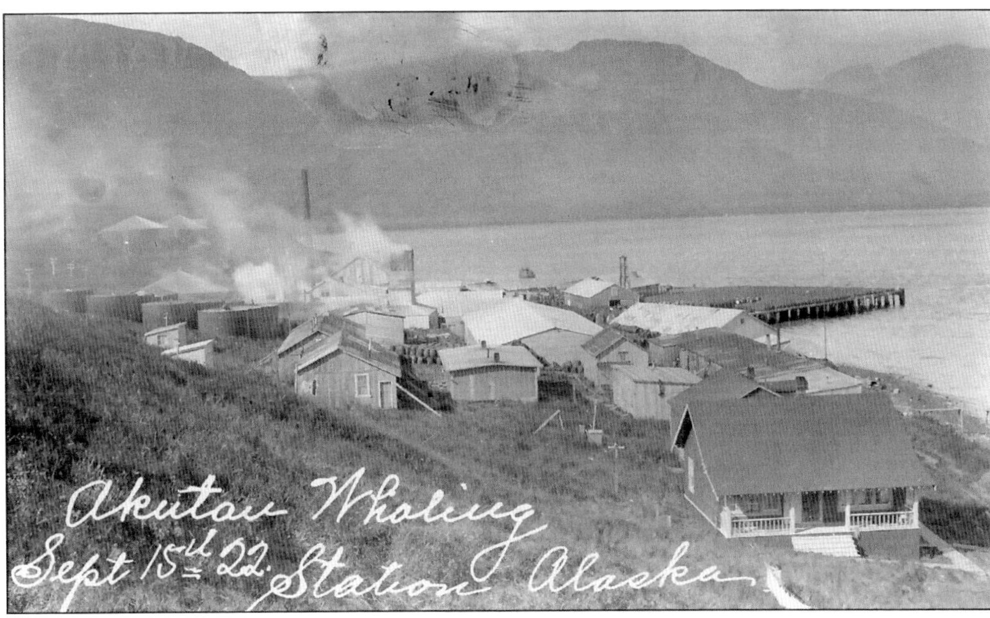

This photograph was taken on September 15, 1922, which would have been at or near the end of the season. Note that there are more buildings in this image than in the previous one, including the manager's building. The manager's house (in the right foreground) has two front doors, which may indicate two residences. The superintendent of the company stayed here during the season. The bark *Fresno* is not present and may have already left for Seattle. The factory seems to be running, but there are no barrels sitting around. (Courtesy of Nore Collection.)

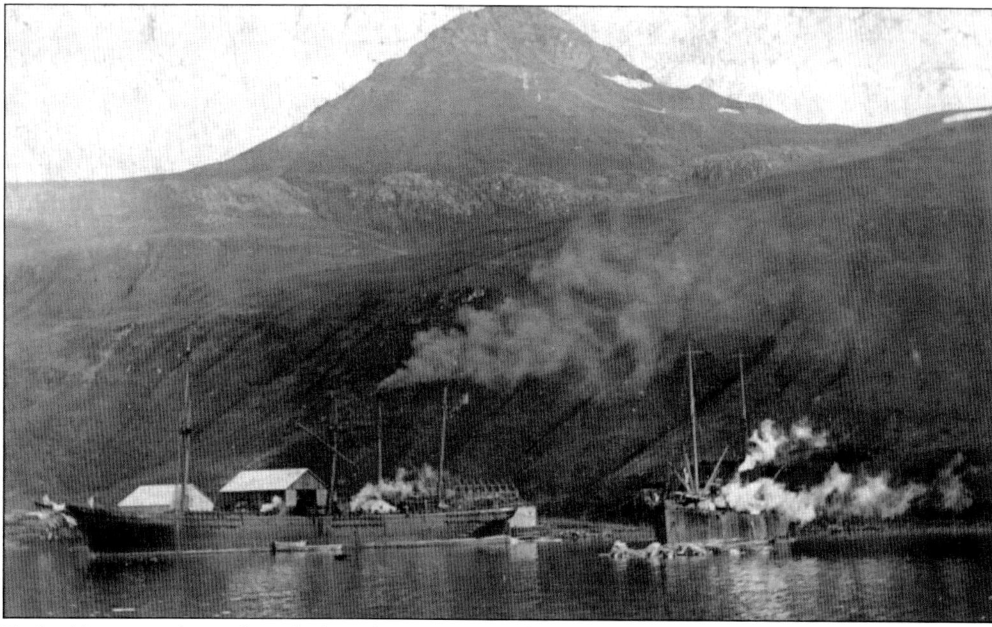

Early Akutan is shown here before the station was built. The bark *Fresno* is on the left. Another ship (unidentified) is at right being used as a boilery. There are five or six whale carcasses tied up around the boiler. The whales have the blubber removed in the water on the side of the ship. Some flensers stand on rigging over the water and cut the blubber from the top.

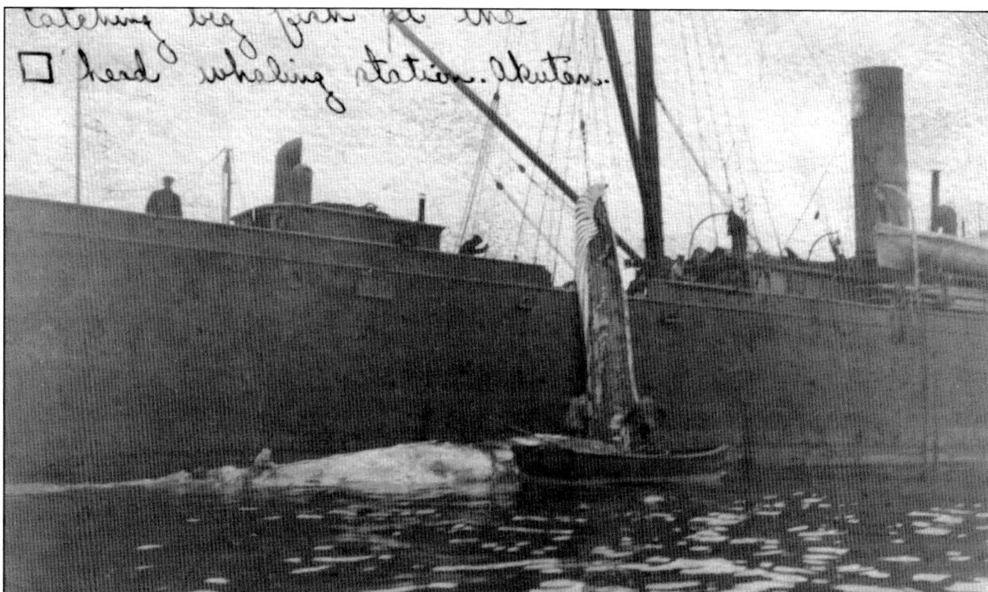

The small dory in the water allows another flenser to work on the whale. The blubber blanket pieces are unrolled around the whale. The whale will be brought into shore for further processing after the blubber is cut off. The ship is used for processing the blubber into oil, which will be barreled for shipment back to Seattle.

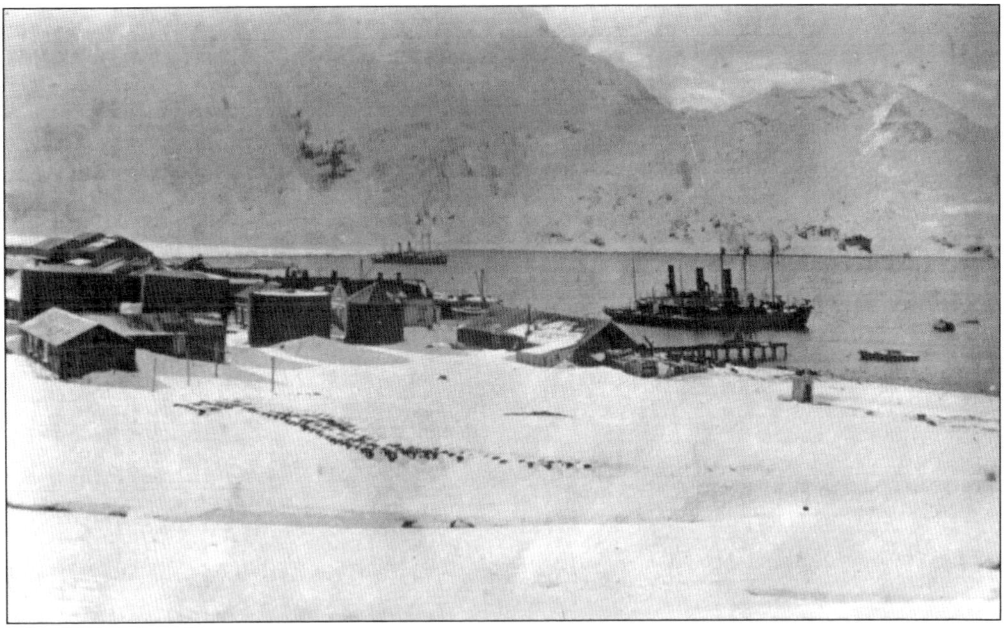

These buildings were the quarters and support buildings for the station workers. The Japanese workers were separated from the other workers by as much distance as possible. Japanese workers had separate quarters due their numbers, which was probably an arrangement agreeable to both the Japanese and American workers.

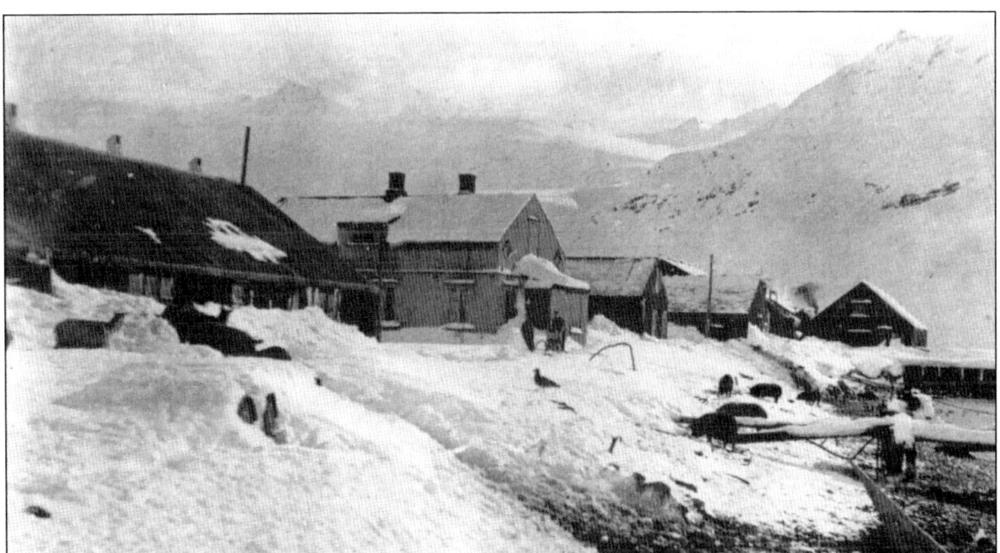

The living quarters at the station show a good complement of snow. Many of the pictures were taken in good weather, but snow, fog, wind, and rain were not uncommon during the whaling season.

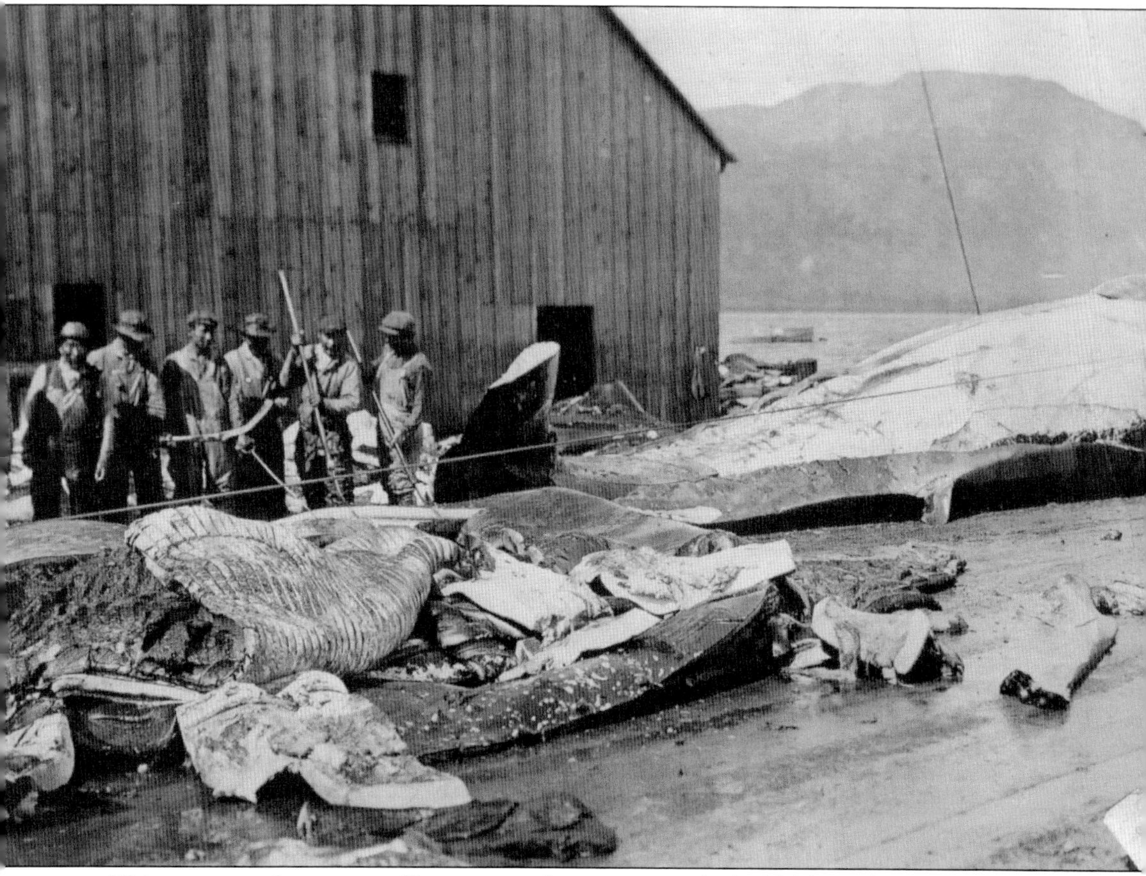

This appears to be a group of Japanese workers posing with the blubber, which is being cut down before being put into a boiler to test the oil. A variety of workers from various countries worked at these whaling stations. The Japanese are the only workers who appeared to have separate quarters from the other workers.

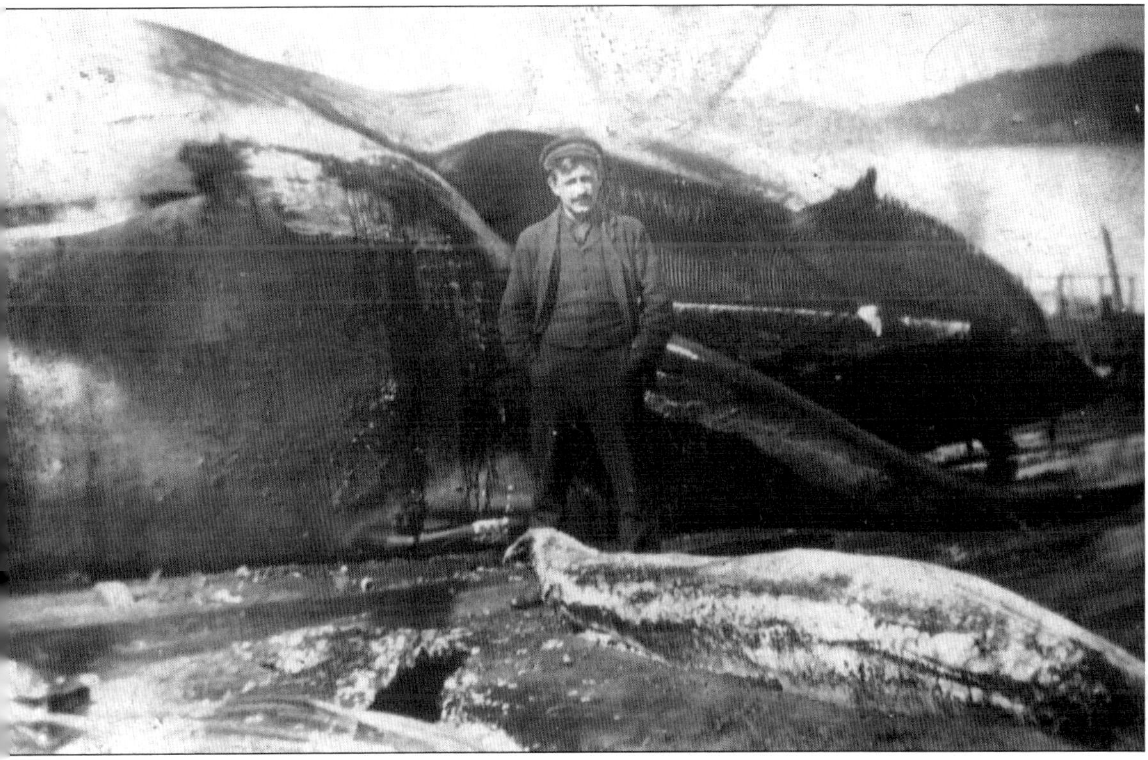

This is a bowhead whale prior to flensing. Note the tall black baleen plates to the right of the unidentified man in the picture. The bowhead has the longest baleen of any whale—it can be 15 feet in length in a mature adult. Scientists once thought the bowhead whale could live to be 200 years old as a result of reading the layers of baleen (similar to reading the rings of a tree to determine its age).

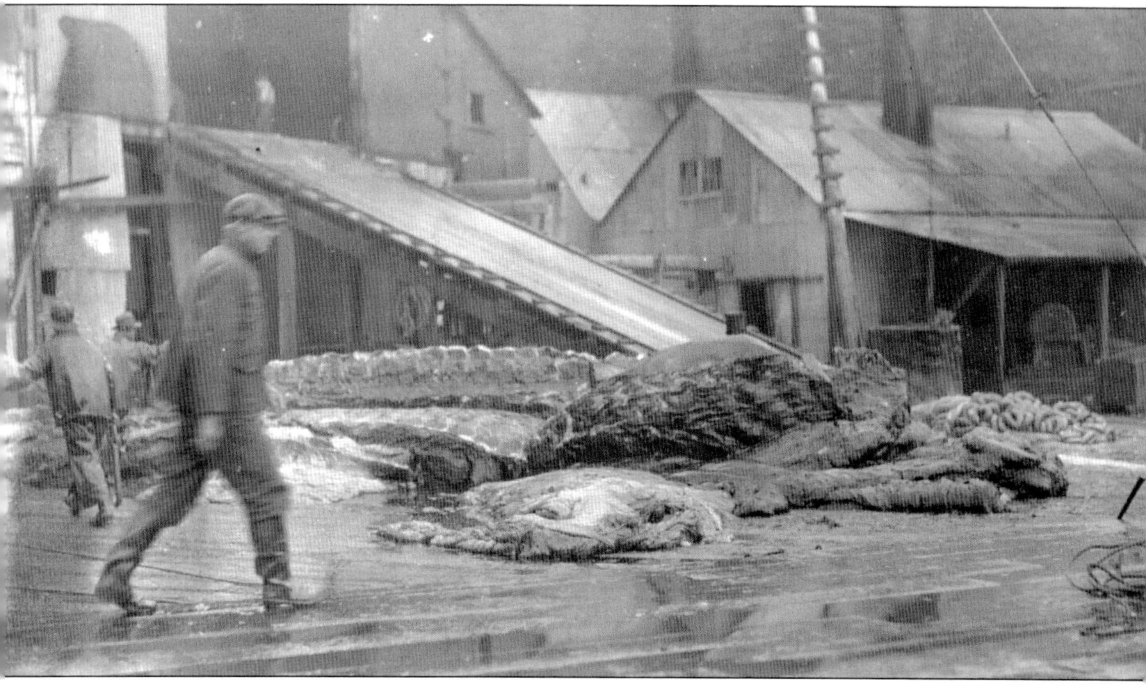

The bones and meat will be taken up the ramp for processing in fertilizer.

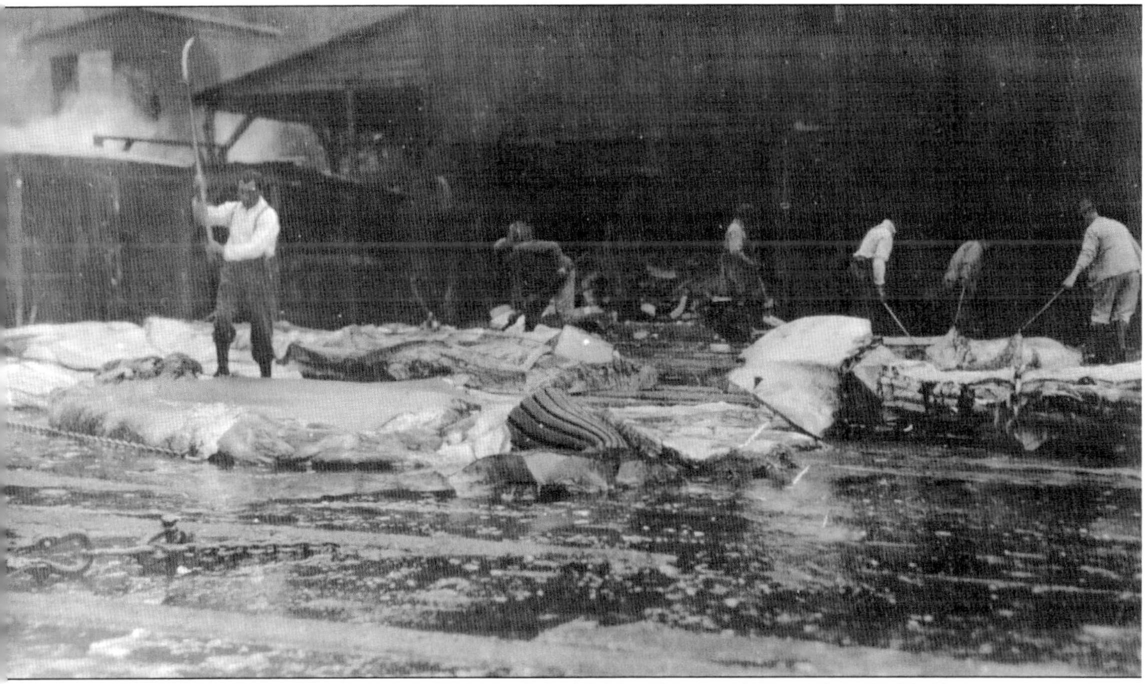

This group of workers is cutting down the last of the blubber from a whale. After the blubber is cut in large blanket pieces from the whale, a team of workers attack it to cut it into manageable pieces for the boilers. After being boiled, the blubber pieces leave remnants similar to pork chitterlings; these are usually thrown into the fire of the boiler.

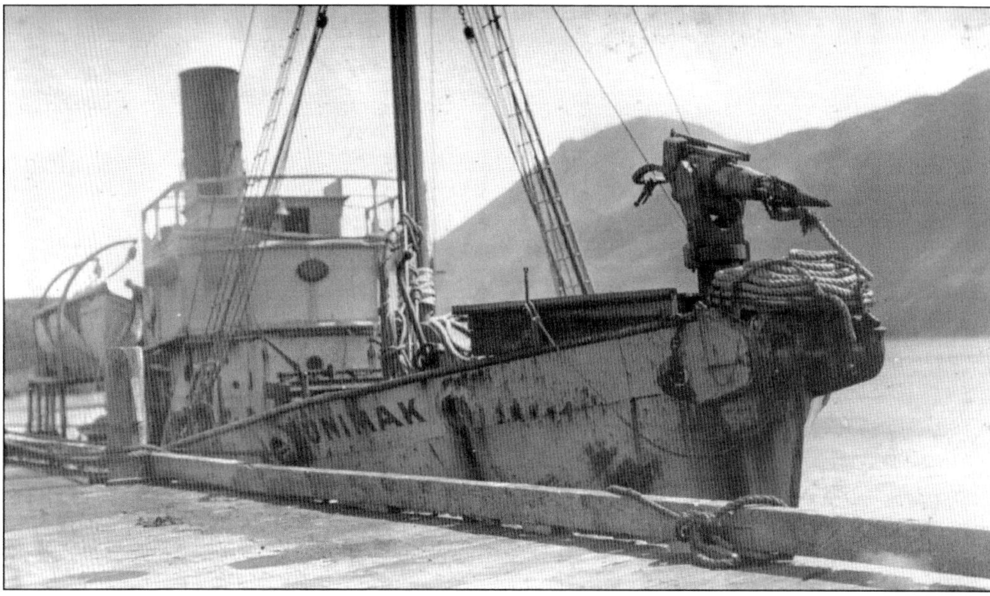

This photograph shows the whaler *Unimak* at the company dock at the Akutan whaling station. The harpoon cannon is ready to go. Many whaling companies do not repaint their vessels between seasons, and whalers can be identified by the rust. Normally, rust does not cause whalers too many problems, but once, when the *Admiralen* was steaming to Akutan, it lost its rudder due to rust and had to be towed into San Francisco for repairs.

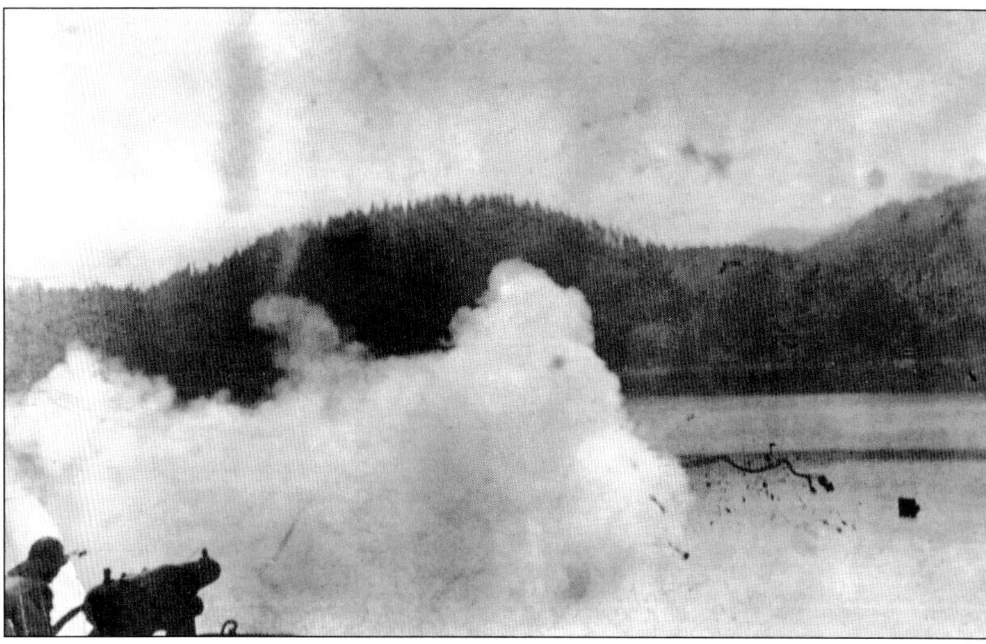

This gunner just fired a harpoon into a whale. There is a lot of smoke—and sometimes parts of the whale—in the air from the explosion. The whalers try to kill the whale with the first harpoon, but this does not always happen. After the whale slows down, the whalers may fire a second harpoon into the whale to kill it.

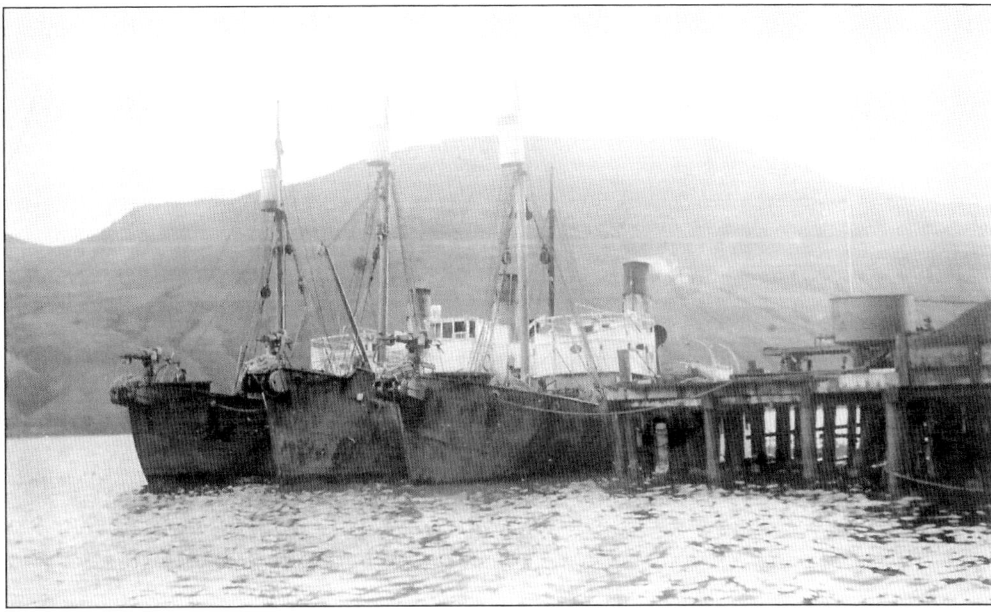

This picture looks to be of the *Unimak* and its twin, *Kodiak*, on the outside, with the *Tanginak* in the middle. The *Tanginak* (formerly the *Tyee Junior*) has a second mast in the rear that the other two boats do not, plus an enclosed cabin. The *Unimak* and *Kodiak* have open cabins. (Courtesy of the Nore Collection.)

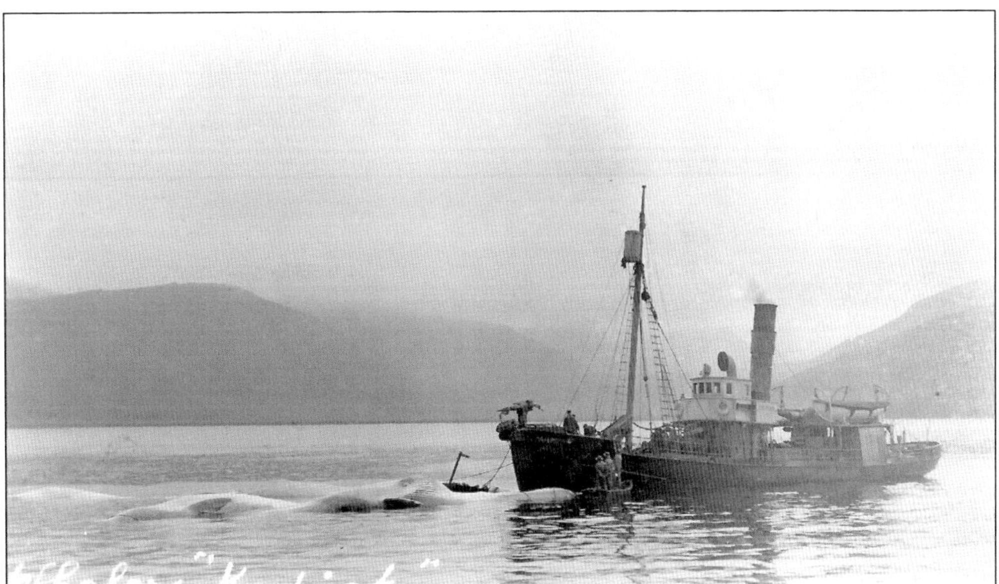

The writing on this image says *Kodiak*, but under magnification, the ship is the *Paterson*. This is a closed-cabin ship with a large air intake over the cabin—these attributes are not present on the *Kodiak*. (Courtesy of the Nore Collection.)

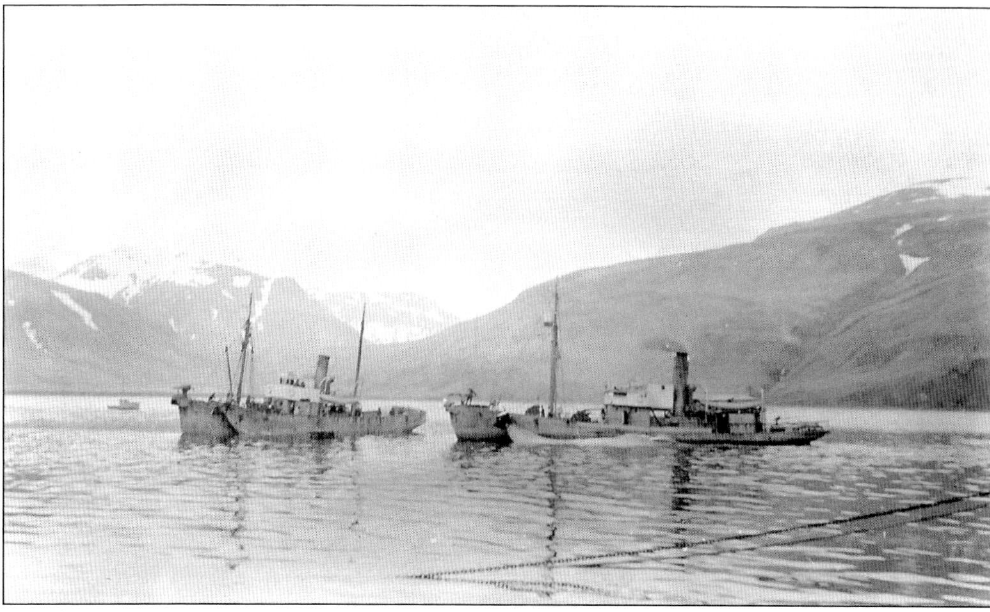

The ship on the left is the *Tanginak*, and either the *Unimak* or *Kodiak*, which are virtually identical, is on the right. Whatever the season, it must have been cold. Note the covering of snow on the mountains, which reaches almost to sea level. (Courtesy of the Nore Collection.)

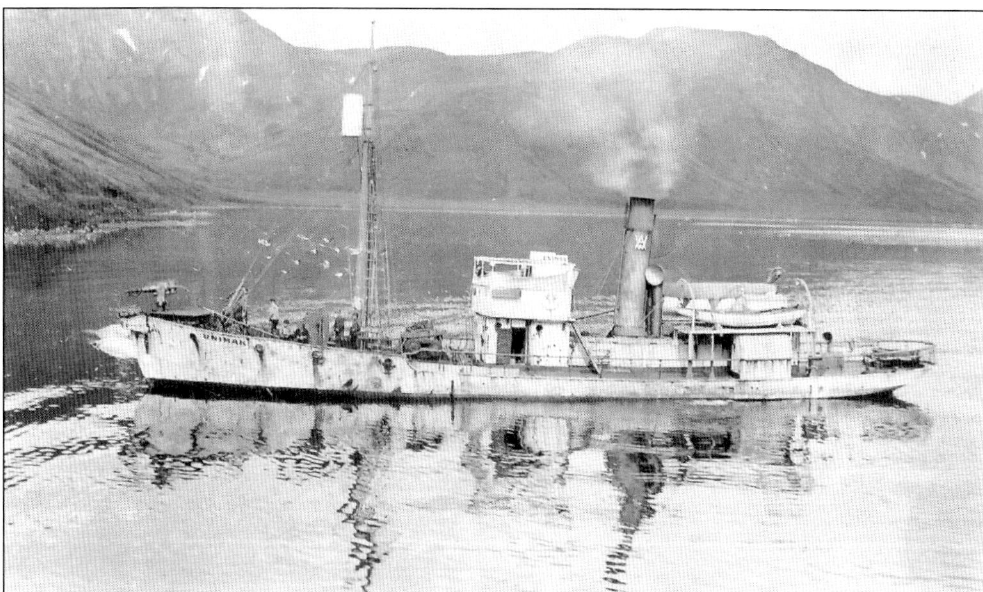

This is a photograph of the *Unimak*, twin of the *Kodiak*. Note that there is no second mast (like the *Tanginak* has) and no enclosed cabin (like the *Tanginak* and *Paterson* have). (Courtesy of the Nore Collection.)

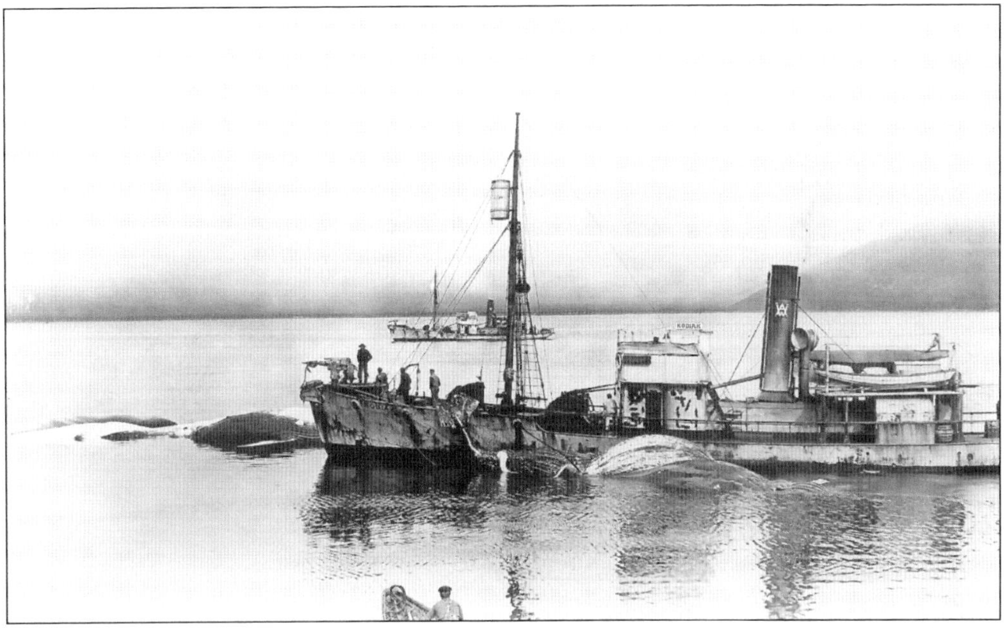

This is a photograph of the *Kodiak* in front with its twin, *Unimak*, in the background. Both are at Akutan with whales to be tied out at the station until they can be taken in for flensing. (Courtesy of the Nore Collection.)

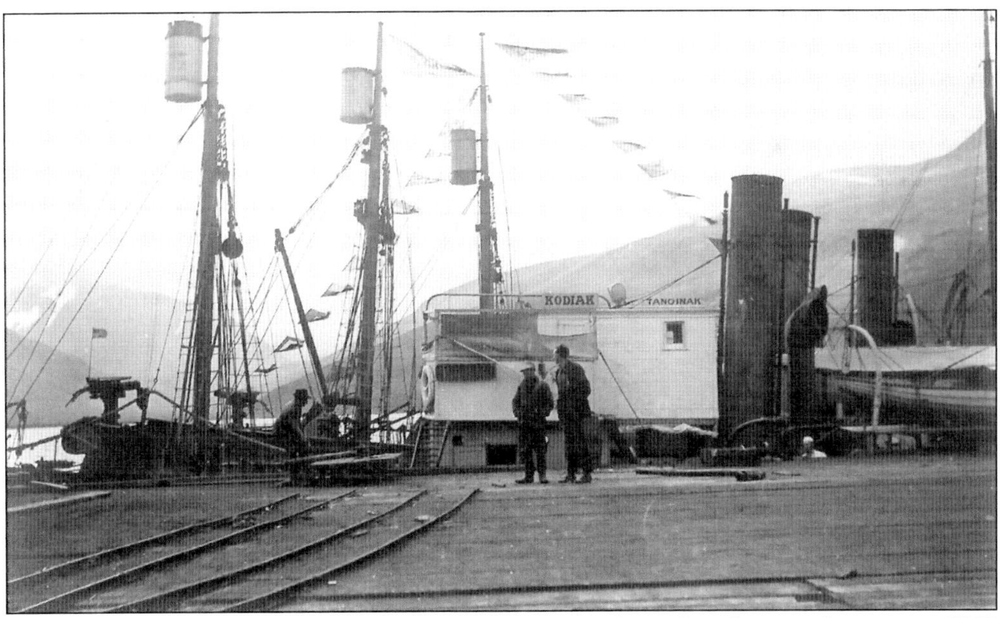

The whalers here are looking good for some occasion. They have flags flying and even a small US flag on a stick on the *Kodiak*'s cannon. Since the boats look like they have a new coat of paint, this could be the start of a new season. The ships here are the *Kodiak*, *Tanginak*, and one other, probably the *Unimak*, which was generally in the company of one of the other two ships.

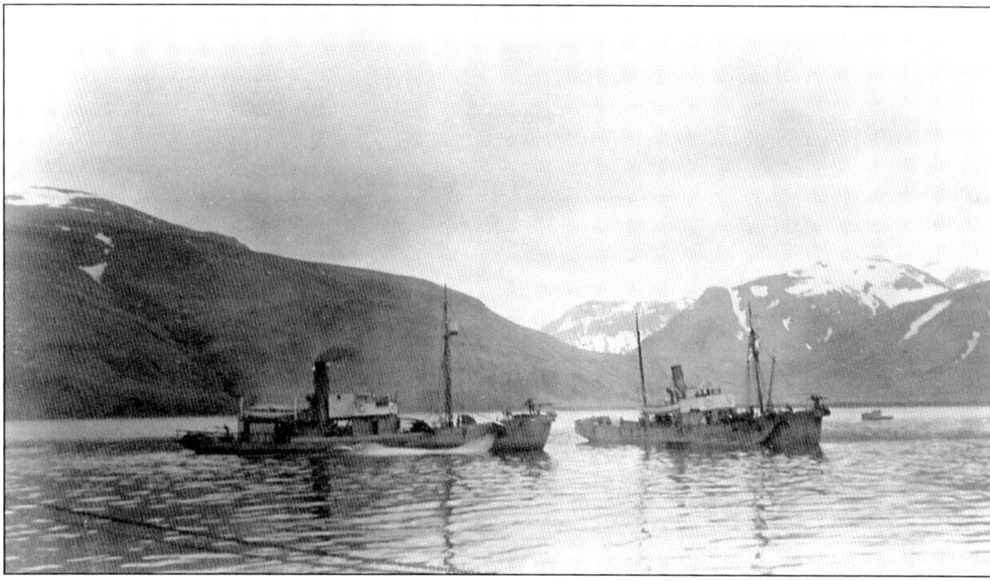

This photograph shows two of the whalers at Akutan delivering whales. The boat at left with the open cabin is either the *Unimak* or *Kodiak*, and the one at right with the second mast and enclosed cabin would be the *Tanginak*. (Courtesy of the Nore Collection.)

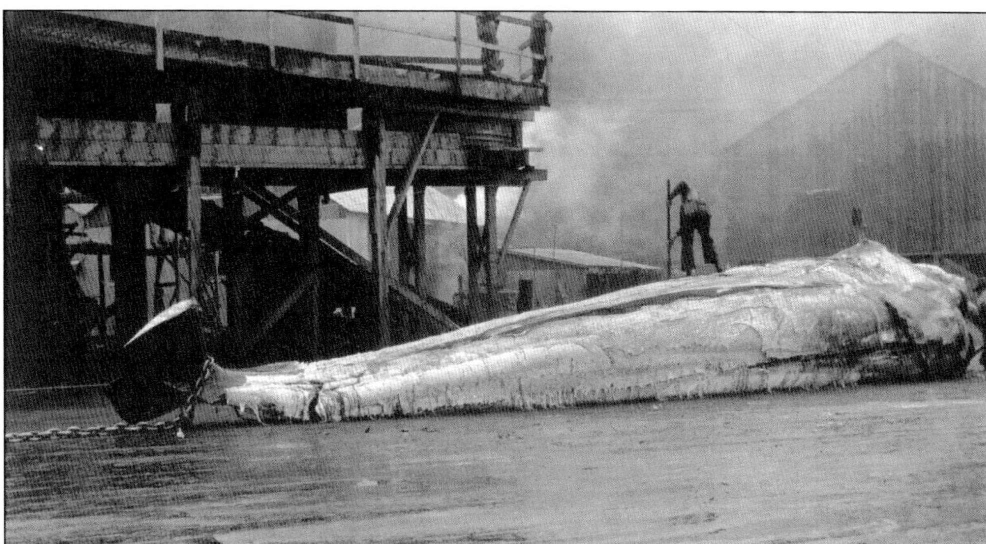

This whale has had the blubber and skin removed and sent to the boiler. The meat will now be removed and sent to the digester; it will eventually be dried and mixed with the ground whale bones to make fertilizer.

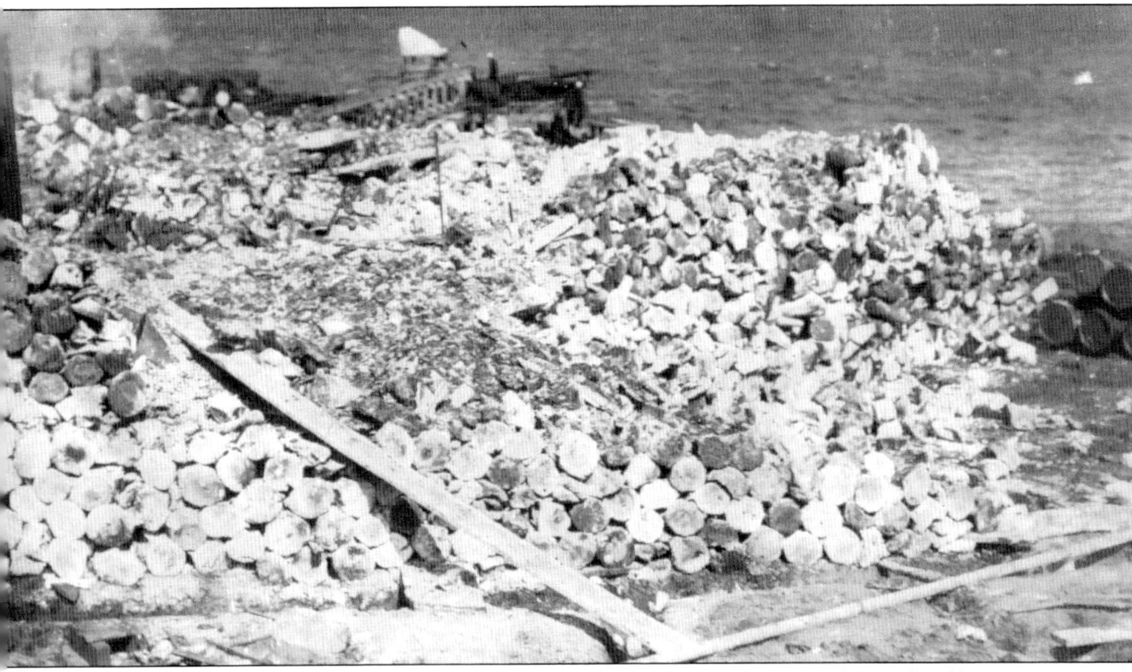

All of the bones from the whales are set aside until they are sent to be ground, usually all at one time. This bone pile reportedly contains the bones of 300 whales, which is about the number of whales taken in one season.

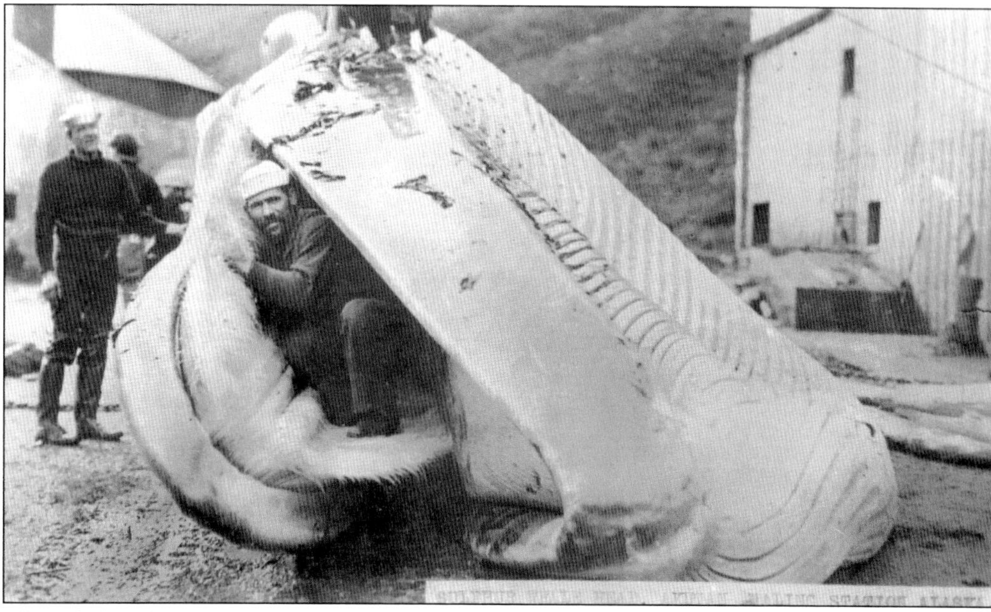

This whale, identified on the image as a sulphur bottom, is more commonly known as a blue whale. The blue whale is the largest animal on earth—the mammal's length can exceed 100 feet. (Courtesy of the Nore Collection.)

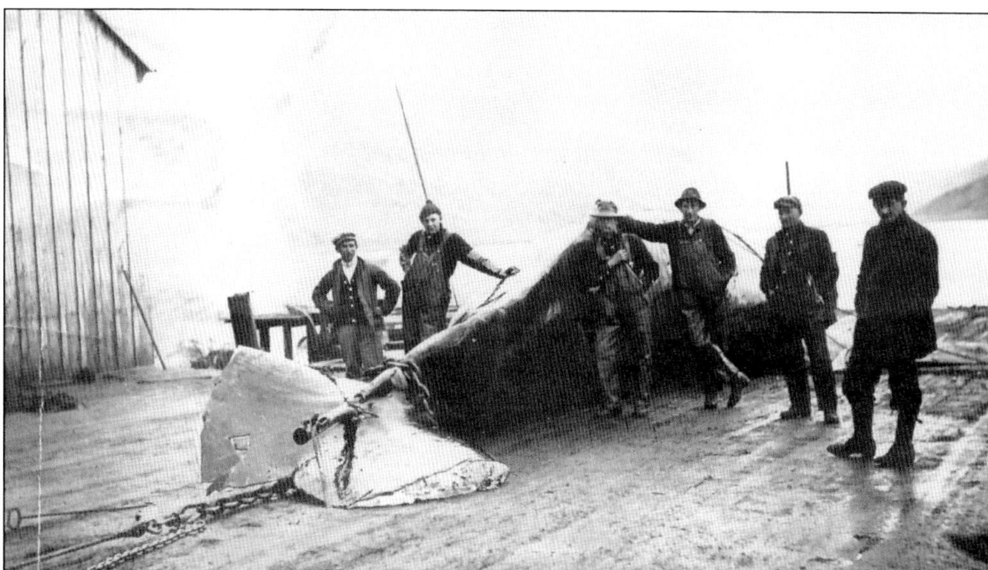

This whale has just been towed in and still has the cable around its tail. It will be dragged by the winch around its tail to the flensing area next to the blubber factory. The tail was cut on both sides. This is done just before the whale is towed in. The whale is secured to the front area of the whaler, and the tail is cut so that when the whale is towed, the tail will not offer resistance. (Courtesy of the Nore Collection.)

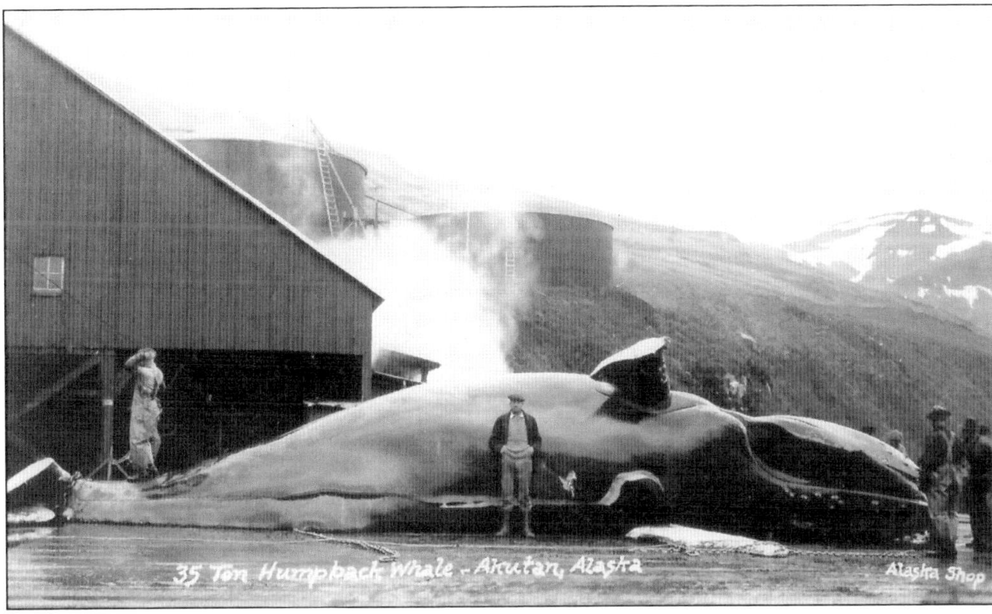

This is a 35-ton humpback whale. The tail flukes and dorsal fins have both been cut for towing. The man in the center is posing with the whale, while the other people appear to be workers, except for one man at far right looking at the whale's mouth. Behind the whale are the blubber factory and the tanks on the hill. (Courtesy of the Nore Collection.)

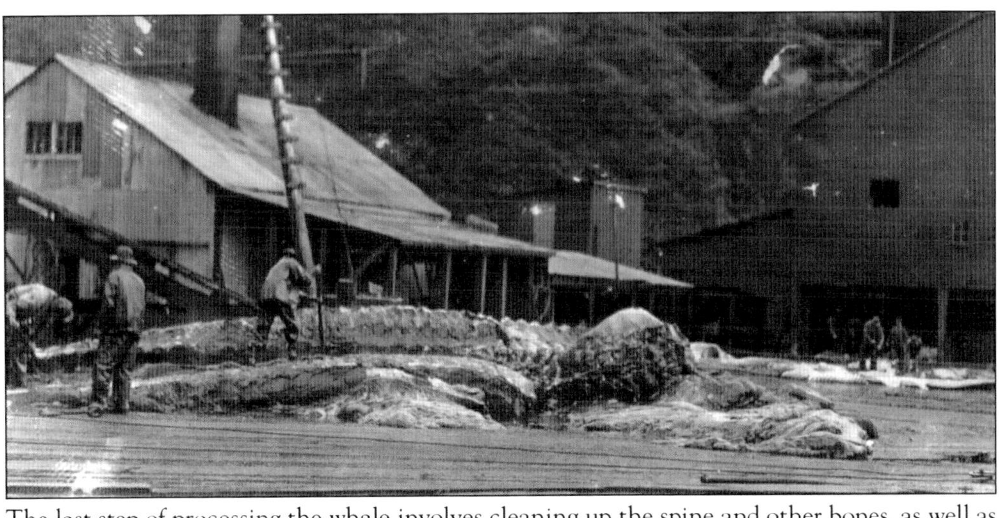

The last step of processing the whale involves cleaning up the spine and other bones, as well as anything else that is left as the process moves along. The little building here may have been used as an office building or quarters. Prior to 1920, the porch was completely enclosed, as it was being used as the engine and boiler room.

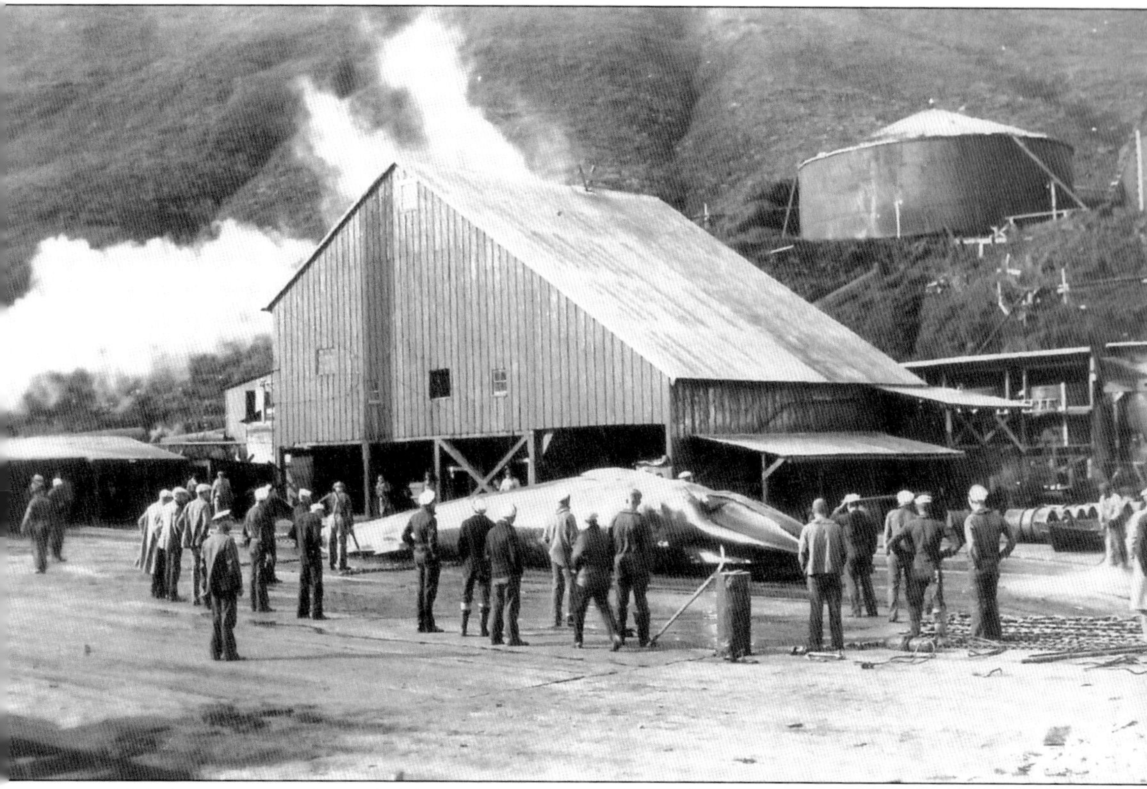

This large group of mostly Navy visitors is watching the flensers start on a large whale. The building behind the whale is the blubber factory, where the blubber is cut down and the oil is then boiled out of the blubber.

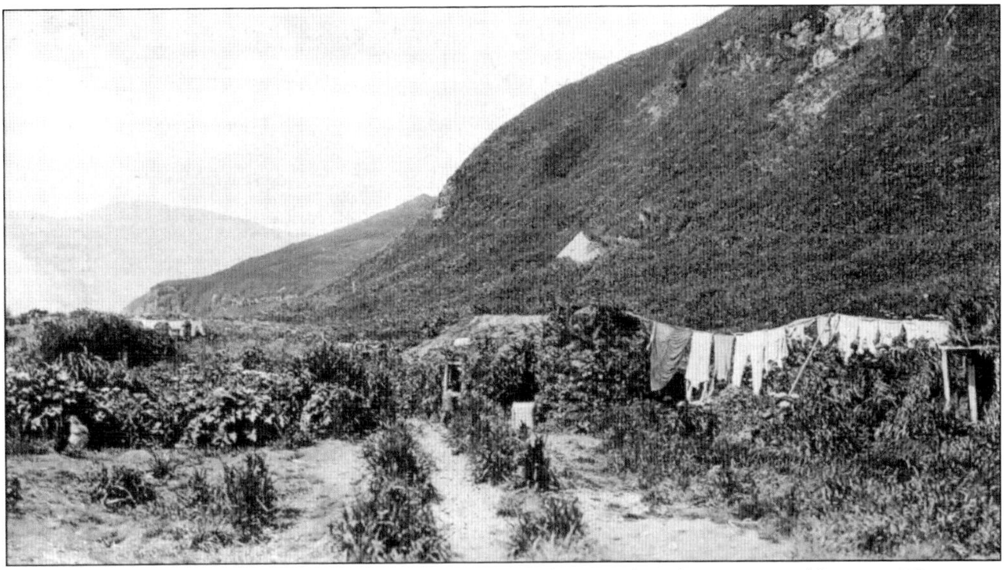

The Aleut village at Akutan shows the obscurity of native barabaras, or sod homes. There are supposedly 25 homes within this village. They are partially underground to provide insulation against the weather.

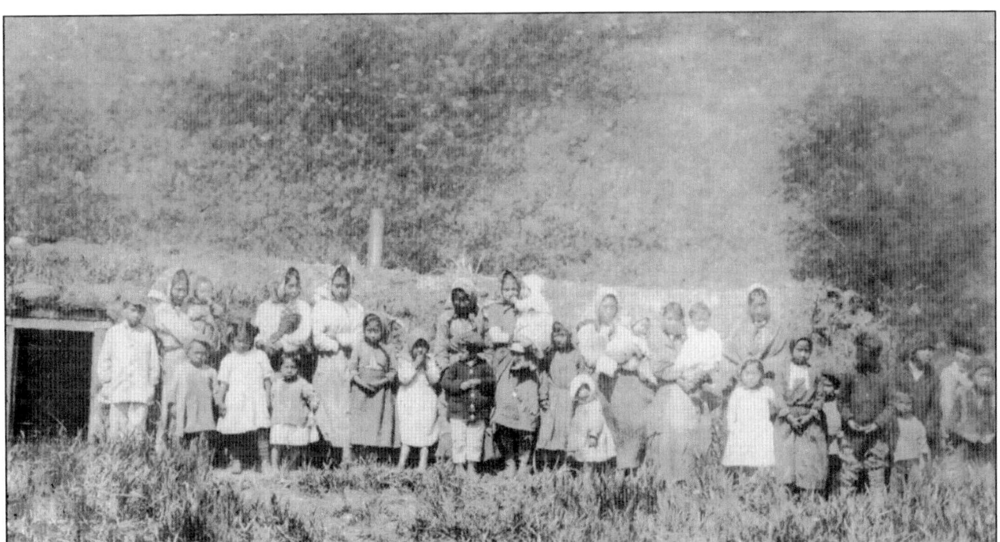

These are women and children from the Akutan Aleut village. The men from this village work at the whaling station. This photograph was taken by E. Lester Jones in 1914.

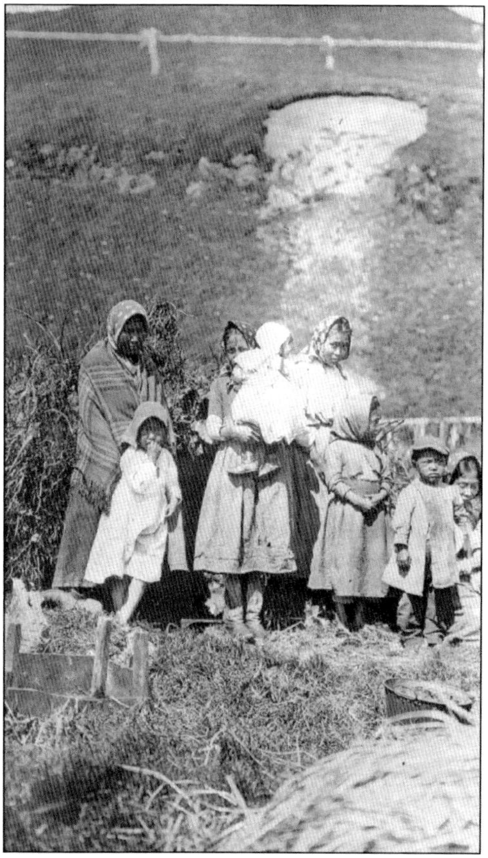

This Akutan Aleut mother and eight children are pictured in front of a barabki, or small type of sod hut. The father was employed at the nearby Akutan whaling station. This picture was taken by E. Lester Jones and was included in the Report of Alaska Investigations in 1914.

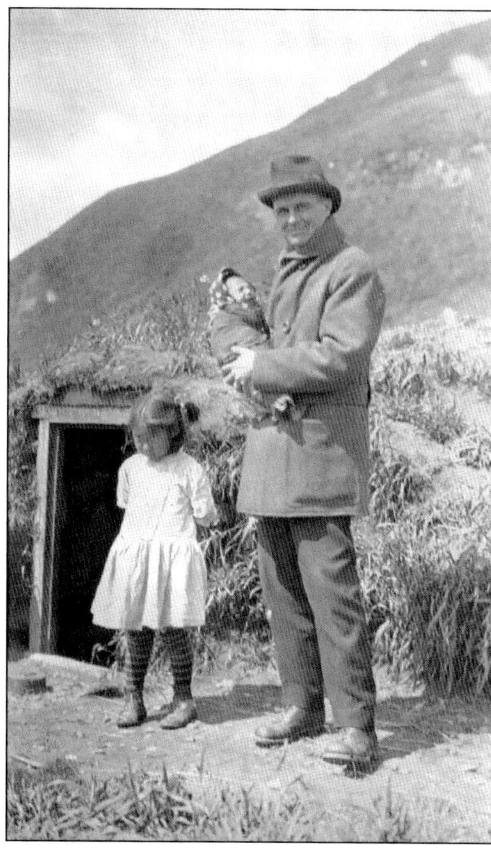

This is a photograph of E. Lester Jones, who visited Akutan in 1914 to take pictures published in the Report of Alaska Investigations in 1914. Some of the photographs used in this chapter were taken by Jones. This photograph was not included in his book and may be published here for the first time.

Six

AKUTAN WHALING STATION, PART TWO

In January 1918, the principal steam whaling concerns of the North Pacific were consolidated through the purchase of the North Pacific Sea Products Company by Victoria Whaling Company interests. The North Pacific Sea Products Company had for several years operated the large whaling station at Akutan, Alaska, with three steam whalers—the *Unimak*, *Kodiak*, and *Tanginak*—and the barge *Fresno* and was owned by Knut B. Birkeland, of Minneapolis, and associates.

Another deal was also closed by the new owners: the purchase of the Pacific Cold Storage Company of Tacoma. This gave them possession of a large cold storage plant as well as the large cold storage steamer *Elihu Thompson*. The *Elihu Thompson* would be operated in connection with the Akutan station, while the Tacoma plant would be conducted as a commercial cold storage plant. A considerable quantity of the company's own products would be handled. This plant was managed by Henry Schupp, who was also general manager for the Victoria Whaling Company.

All of the affiliated companies gave considerable attention to producing and marketing whale meat for food, as they had facilities for taking the species of whales used for this purpose, and the Victoria and American Pacific companies were the pioneers in this field on the Pacific.

Some $60,000 was spent in 1918 on further improvements at Akutan, and the Sechart (Canada) plant was dismantled and moved there. A new cold storage plant, a dryer, and an oil tank with a capacity of 25,000 barrels of oil were also installed. Whale meat was frozen and shipped to Seattle on the *Elihu Thompson*. The Akutan whaling fleet sailed from Seattle in mid-May and was followed shortly by the *Fresno* in tow of the *Elihu Thompson* with a full crew and supplies.

In June 1918, the *Elihu Thompson* sailed from Puget Sound with a cargo of meats and perishable goods bound for Nome and St. Michael via Ketchikan. It stopped at the Akutan whaling station on the way out and on the homeward trip to bring a large quantity of whale meat to the distributing plan at Tacoma. In July, the whaler *Paterson* was transferred to the company's station at Akutan. The tender *Gray* arrived at Tacoma with a full cargo of whale oil from Akutan for the company's new oil storage plant. The bark *Fresno* left in early July, the last time it would be required for processing whales, since the Akutan plant could now operate independently.

In early November 1918, the steam whalers *Kodiak*, *Tanginak*, *Unimak*, and *Paterson* arrived at Seattle from the North Pacific Sea Products Company's station, which closed the most successful season in company history with a total kill of 310 whales, 111 of which were taken by the *Kodiak*. In 1919, the company recorded a catch of 1,025 whales and produced 35,000 barrels of oil, as well as a large quantity of guano and frozen whale meat. In 1920, it yielded 32,800 barrels of oil and 23,000 sacks of guano.

Henry Schupp, president of the North Pacific Whaling Company, stated that the market for frozen whale meat was steadily growing and that canned meat was well received in a number of markets. He predicted a large volume of foreign business could be had if buyers could secure import permits. The production of canned and frozen whale meat was strictly carried on during the World War II years. The public's interest in whale meat waned when beef and other meats came back on the market after the end of hostilities.

"Intense propaganda attempted, albeit in vain, to introduce whale meat for domestic consumption on a par with beef and mutton," write Tønnessen and Johnsen in *The History of Modern Whaling*. The company closed its public cold storage plant at Tacoma and began to transfer the meat from Akutan direct to the cars from the cold storage steamer *Elihu Thompson*. Schupp now made his headquarters in the L.C. Smith building in Seattle.

The company was considerably hampered in 1919 by an epidemic of influenza, which was severe among the Aleutian Island natives and prevented the catch from being even larger.

Tønnessen and Johnsen further assess the impact this had on the industry:

> Although 1915 through 1918 provided a nice margin, liabilities and repayments of loans proved such a burden that it was never possible to pay any dividend. The end of the boom period in 1919 not only entailed a disastrous fall in prices, but also made it impossible to market the oil in the United States. In 1921, most of the two previous years' production was still stockpiled and finally had to be sold at a loss. When the company went bankrupt in 1922, shareholders and creditors suffered a loss of 68,680 pounds, the biggest disaster to date in the history of Norwegian whaling. This terminated Norwegian westward expansion of whaling in the northern hemisphere, a venture which had ended with a crash on both sides of the American continent. Others, however, continued where the Norwegians had left off.

In September 1925, Capt. Fred Peterson suffered severe lacerations to the arms as the result of being struck by a bomb fragment after his harpoon had pierced the body of a whale and exploded. He was able to recover from his wounds.

After the closure of the Bay City station in Washington State, the whalers *Moran* and *Aberdeen* were sent to Akutan. Whaling was carried on at two stations in 1926, Akutan and a new station built at Port Hobron, Sitkalidak Island, about 70 miles from Kodiak.

In 1927, whalers working out of Akutan were the *Unimak* (Capt. N. Thorvick), *Kodiak* (Captain Schroeder), *Westport* (Captain Reed), and *Paterson* (Capt. Peter Oness). Ernest Kugler was in charge at Akutan.

Akutan was shut down for the 1931 to 1933 seasons, all concerns having decided to withdraw due to the low price of whale oil. The company was unable to secure contracts for whale oil at prices which left any possibility of breaking even. The plant at Akutan was used by the Northwest Herring Company for its Aleutian herring salting.

In 1933, none of the North Pacific whaling stations operated for the first time since 1930.

The Akutan station in 1934 was in charge of E. Kugler and was served by four killers: *Umiak*, *Westport* (Harold B. Kristensen, captain and gunner), *Kodiak* (Peter Oness, captain and gunner), and *Paterson* (N. Thorvick, master, and Frank Christensen, gunner).

In June 1934, as reported in *Pacific Fisherman* that year, Capt. Harold B. Kristensen, master and gunner of the whale killer *Westport*, was killed during the season when the recoil of a whale gun he had fired carried away the mounting and allowed the piece to fall upon him. Command of the *Westport* passed to M. Schroeder, the mate.

On September 15, 1934, the 95-foot steel killer boat *Westport* of the American Pacific Whaling Company fleet struck Bight Reef off Akutan Island and was a total loss. The crew, headed by Capt. N. Schroeder and Gunner O. Lie, was taken off by the Coast Guard cutter *Daphne* after two men made shore and walked to the Akutan whaling station to report the accident. The wreck was inspected by Coast Guard officers, who found the vessel damaged so extensively as to make salvage impossible. The crew of the *Westport* was brought by the cutter *Chelan* to the Akutan station.

The Akutan and Port Hobron stations of the American Pacific Whaling Company were idle after 1939. While whale oil was important to the wartime economy, the industry in Alaska yielded several ships to the military and closed operations in the interest of national defense.

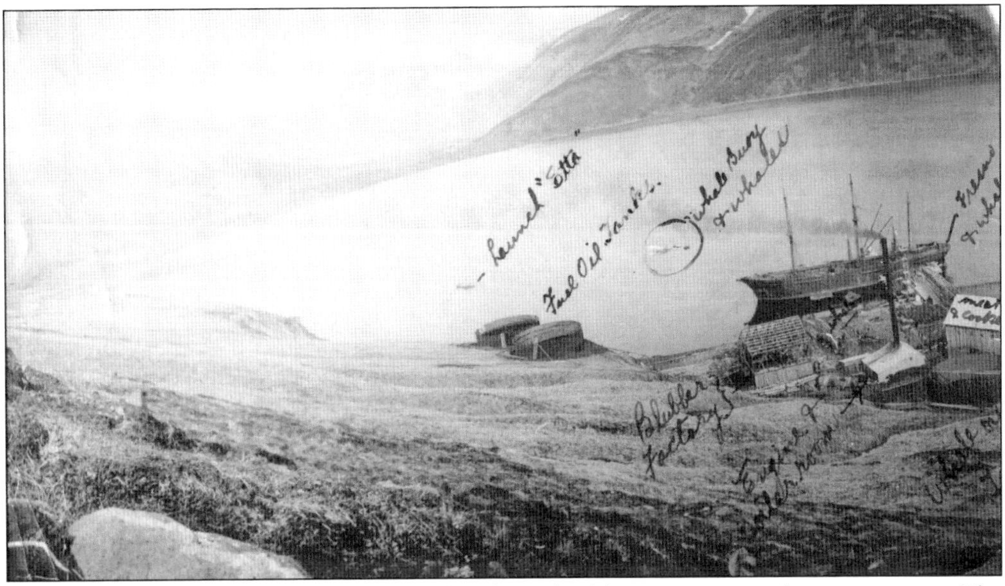

This picture is one of two overhead shots that show the Akutan station under construction. The construction to upgrade Akutan was started in 1914 under new management. Two fuel oil tanks are built, but the five product tanks have not yet been constructed. Also, the construction of the blubber factory has started, but the roof is not yet complete. Some whale carcasses are waiting for processing, and the bark *Fresno* is still being used. This photograph dates to earlier than 1918.

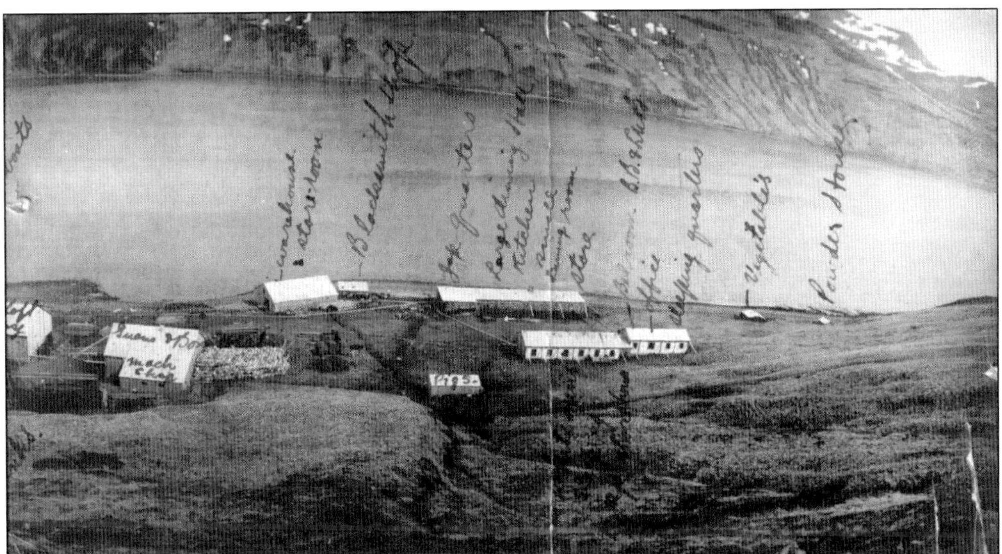

This second photograph shows the newly construction station, with the Japanese sleeping quarters next to the cafeteria and other sleeping quarters far to the right. A little farther to the right is where the station manager's cabin would be constructed in the future. There is a dynamite building far to the right, probably where the charges were created for the harpoons and cannons.

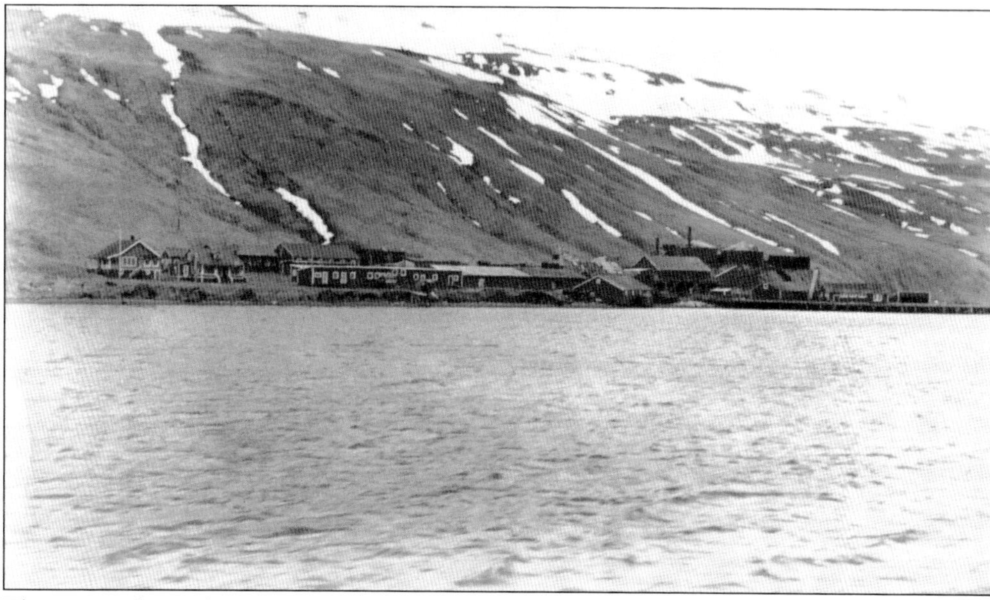

This image shows the approach to Akutan, which was updated some time prior to 1922. Most if not all of the buildings have been erected by this time. Also, the buildings have had exterior trim and window moldings installed. The window trim and moldings shown here are present in many later pictures.

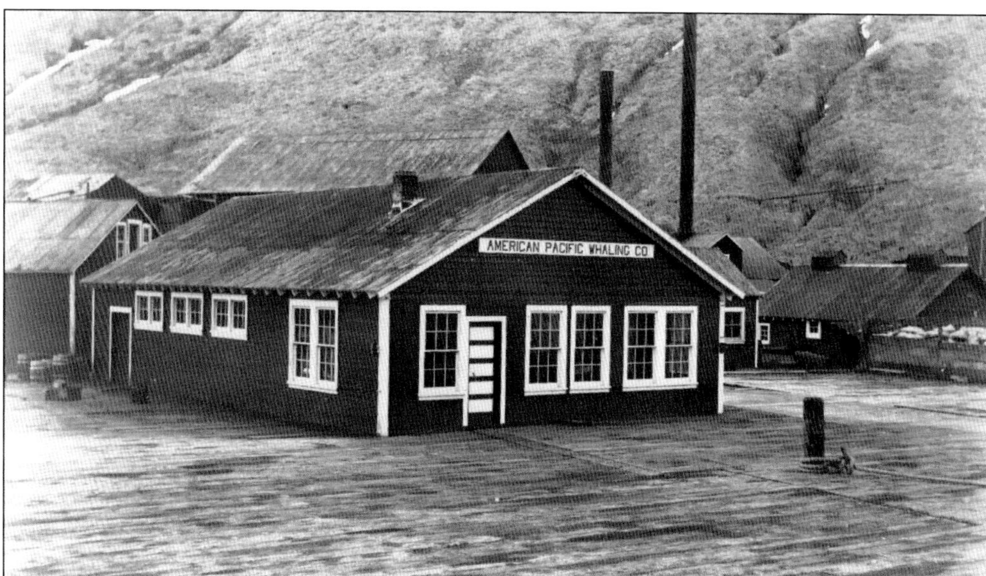

The American Pacific Whaling Company office building was constructed between 1914 and 1922. This office is approximately in the middle of the station, with the factory part of the station to one side and the living quarters and non-factory buildings to the other side.

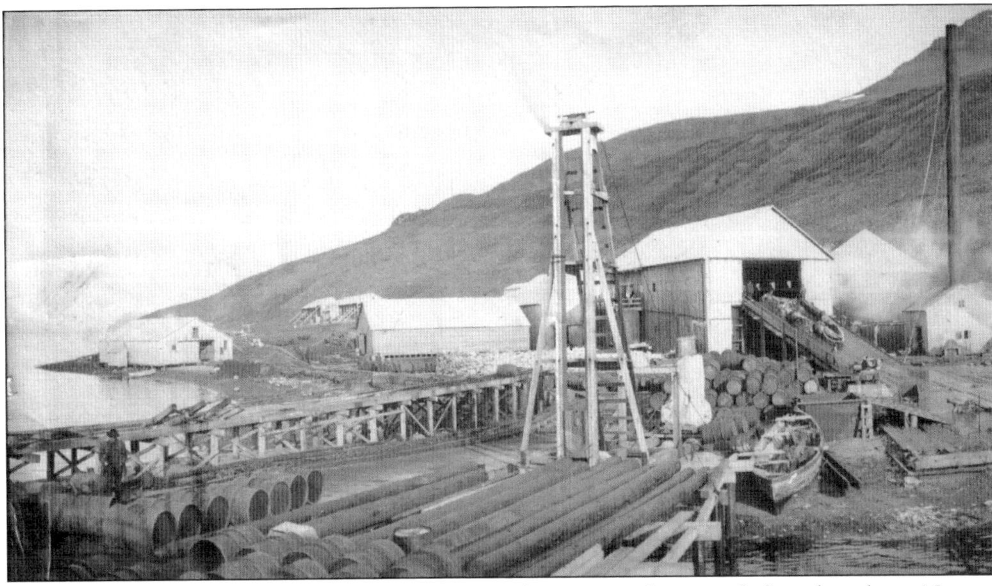

The building at right with the whale carcass on the ramp is the meat loft and cookery. Next to it is the bone pile. The building at left is the warehouse and storeroom. The purpose of the front building in the center is unknown; it was built after the aerial pictures (shown on page 81) were taken. The two buildings in the back are residence quarters.

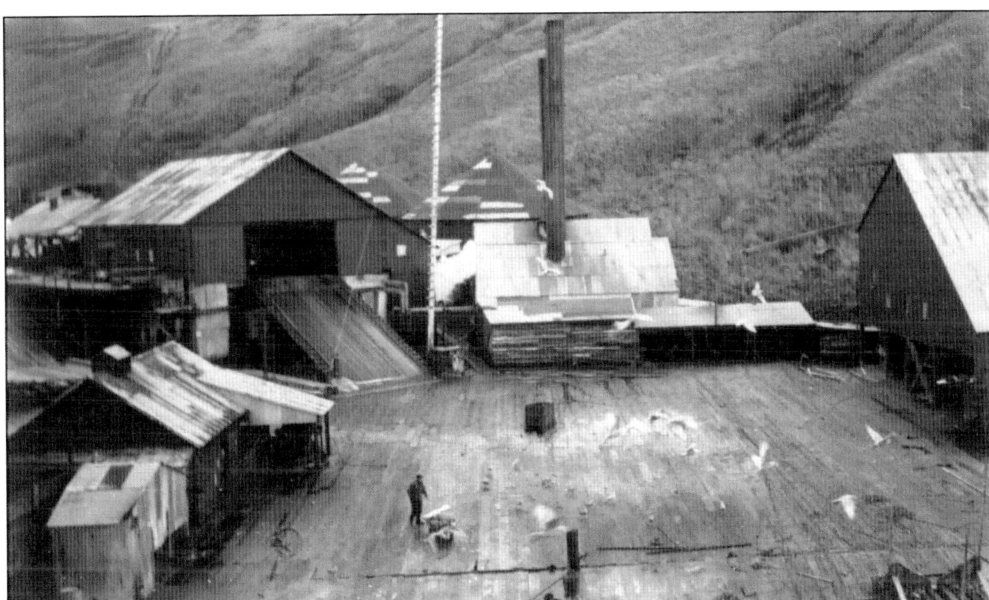

This photograph offers another look at the meat loft and cookery. The blubber has been removed at the first operation before a carcass goes this far. The building in the center was the engine and boiler room; in this picture, it has a porch facing forward that has been enclosed. The side of the blubber factory is partially visible at right.

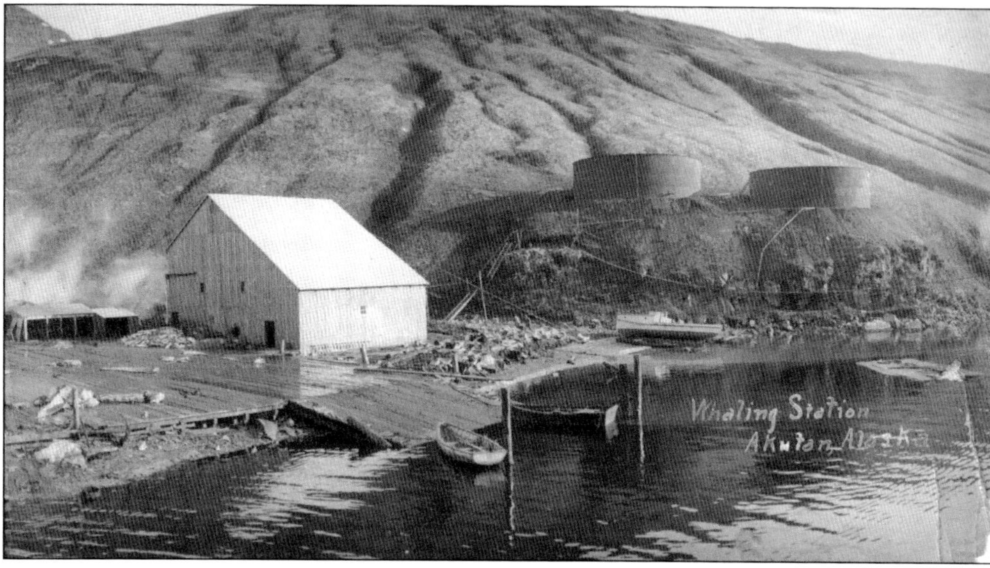

The building at left is the completed blubber factory, which was constructed during or after 1914. The two fuel oil tanks on the right were also built during or after 1914. To the left of the blubber factory are two small shed-type structures. The viewpoint of this picture is approximately where the *Fresno* would be anchored, and it could have been taken from the deck or possibly the mast (to get altitude for this angle).

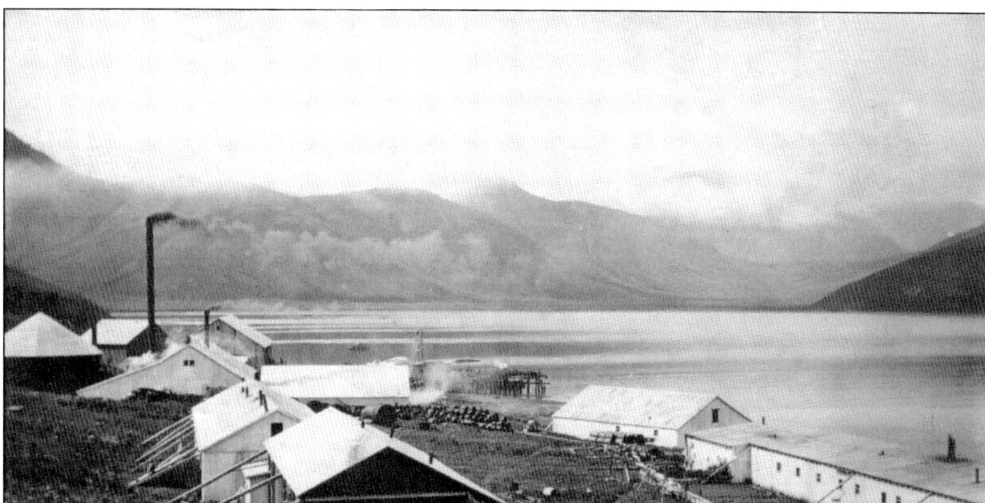

This picture shows the residence quarters in the foreground, and at right are the office, cafeteria, and Japanese residence. Toward the factory are the warehouse and storeroom with a blacksmith shack next to it. In the center is what appears to be a stack of drums used for shipping oil.

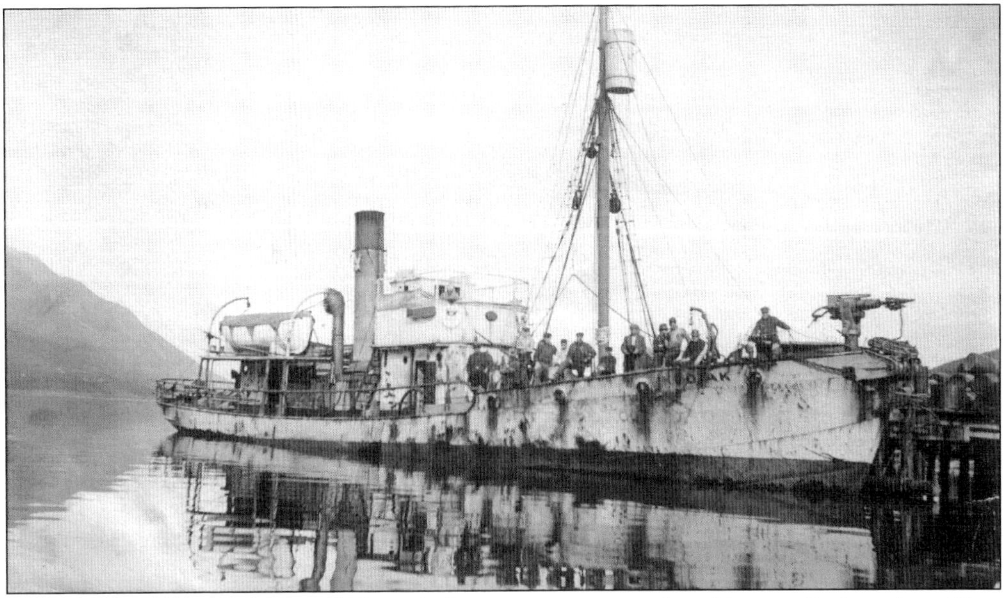

This is a photograph of the *Kodiak*, a whaling steamer assigned to Akutan. On board are 12 to 13 men, which was about the complement of the crew. Whalers were notorious for being very rusty. This is probably because in the season, they were out searching for whales for most of the available time.

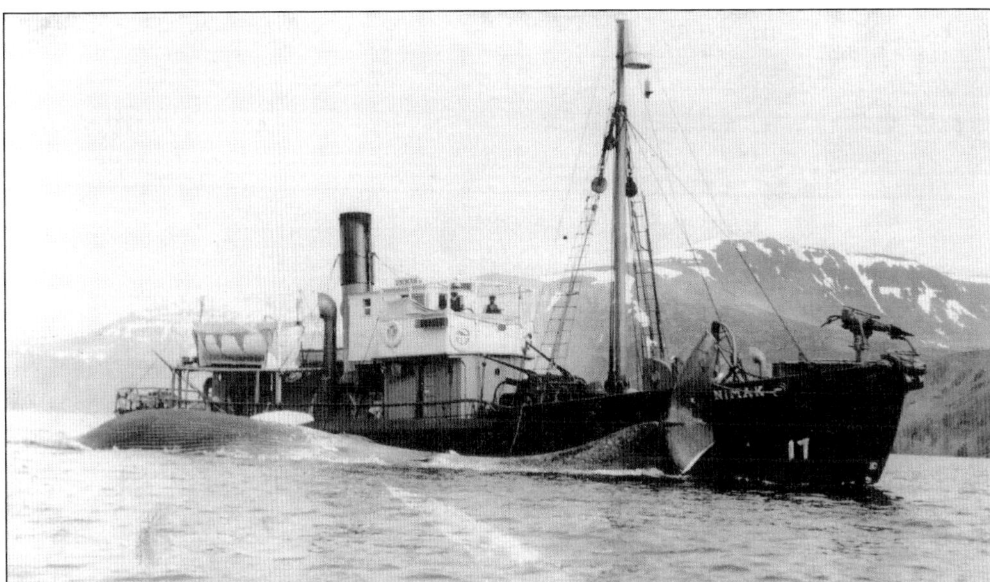

This photograph shows the *Unimak*, sister ship of the *Kodiak*. In this picture, it is nicely painted, which might have been for the start of the season. Note that the whale's tail flukes have been cut off; this keeps the whale from interfering with ship's navigation. As on the *Kodiak*, the cabin is open and there is one mast.

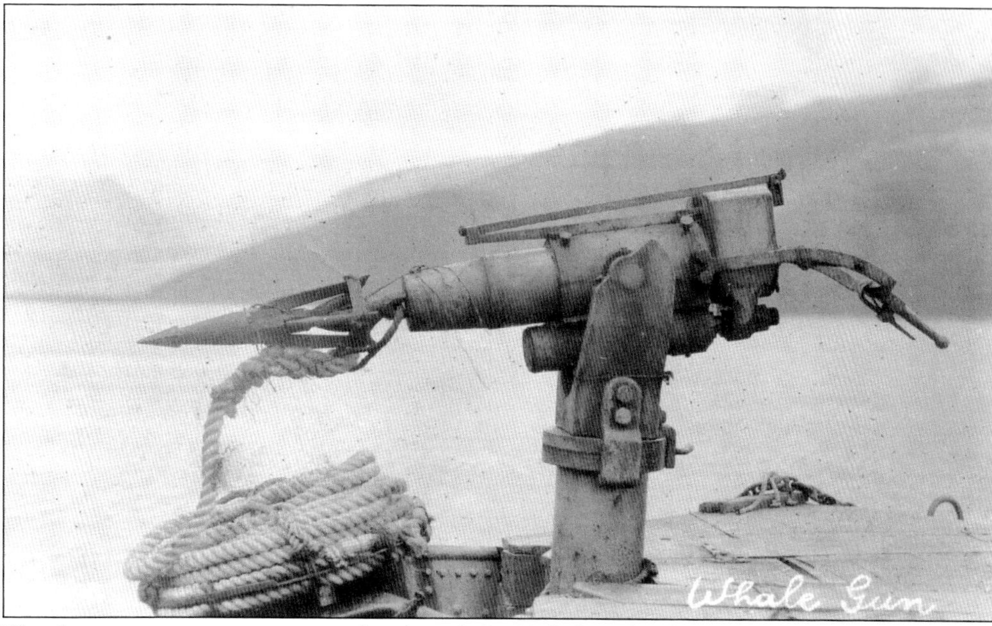

This harpoon cannon appears to be aboard the *Kodiak*. It has been loaded with the harpoon and dynamite charge. There is also a charge in the end of the harpoon that will detonate a few seconds after impact. When the harpoon is fired, it will take the coil of rope with it. The harpoon will extend the barbs after it goes in the whale and should hold tight. When the dynamite charge goes off in the whale, it is designed to kill if the harpoon lodges near a vital spot. If not, more harpoons may be necessary.

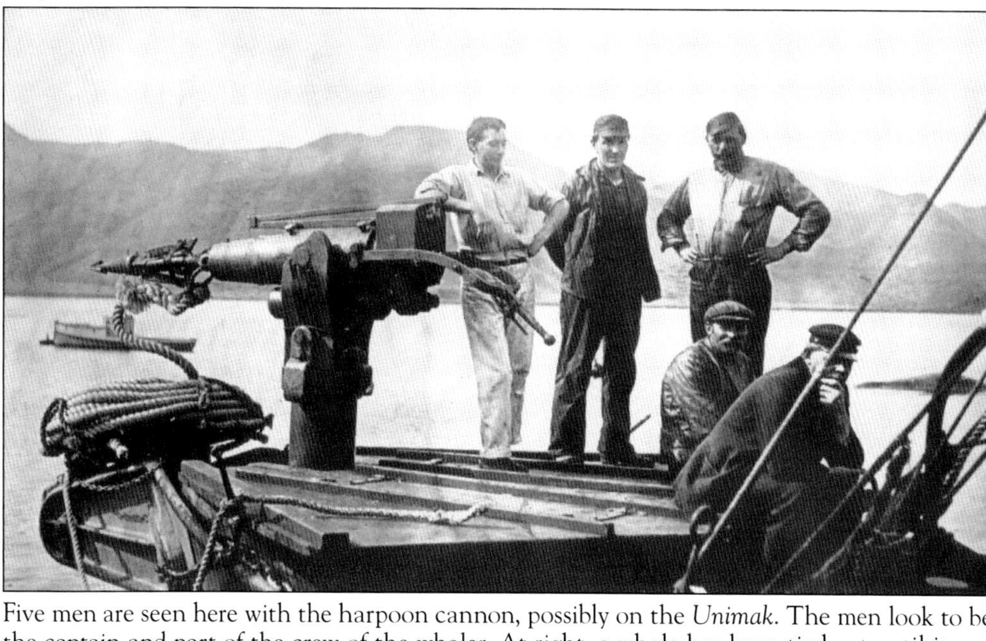

Five men are seen here with the harpoon cannon, possibly on the *Unimak*. The men look to be the captain and part of the crew of the whaler. At right, a whale has been tied out until it can be brought in for processing. After the gun is fired at the whale, as the whale dives, it will pull out more rope. The rope is attached to a winch that works to maintain tension and draws the rope up as the whale tires.

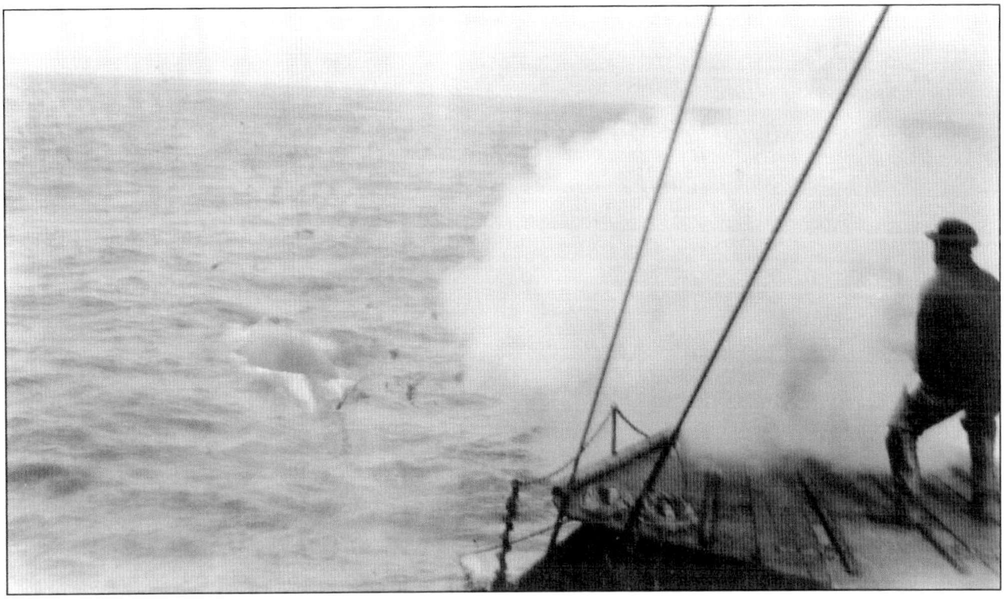

These whalers are firing a harpoon from aboard a killer boat off the coast of Alaska in 1915. A killer boat is a harpoon-mounted boat designed for killing whales. (Courtesy of University of Washington Libraries, Special Collections, UW4856.)

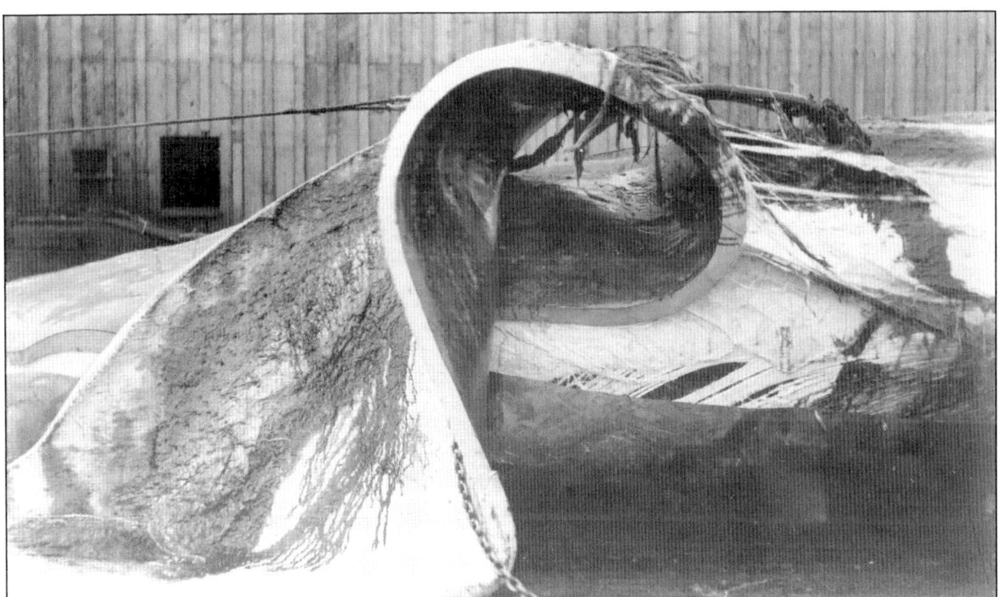

At what is likely the whaling station at Akutan, Alaska, men are flensing a blue whale in 1915. (Courtesy of University of Washington Libraries, Special Collections, UW 33490.)

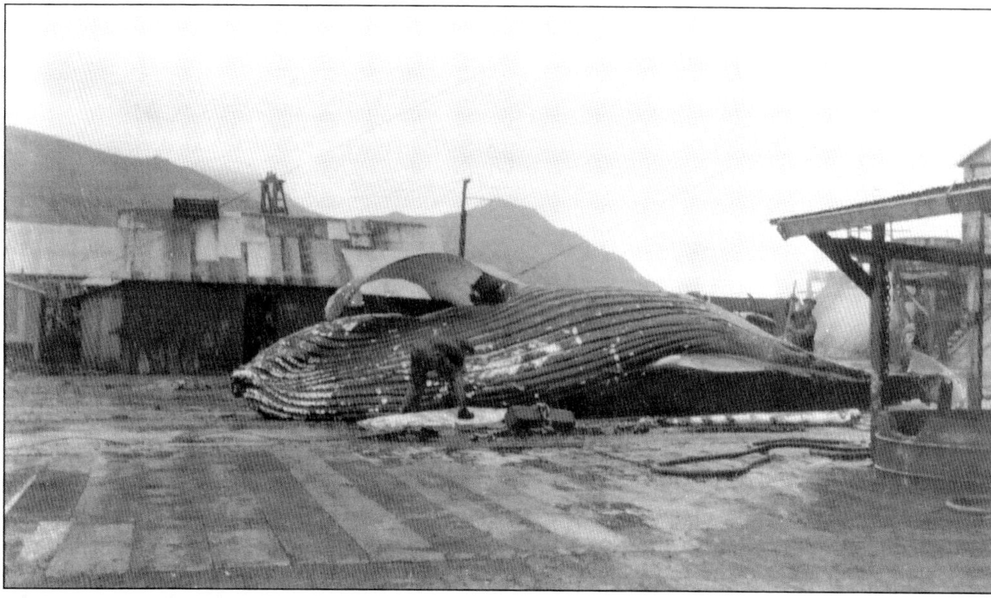

This appears to be a humpback whale. It has had a cable placed around the tail and was dragged by steam winch to the blubber factory. The blubber factory is the building behind the whale. Fuel oil tanks have been added to the right. The head flenser, the man holding the long-handled flensing knife, is making the initial cuts into the blubber. This cut will go all the way down the whale, and then a winch will be connected to the blubber by the head end in order to pull off a blanket piece.

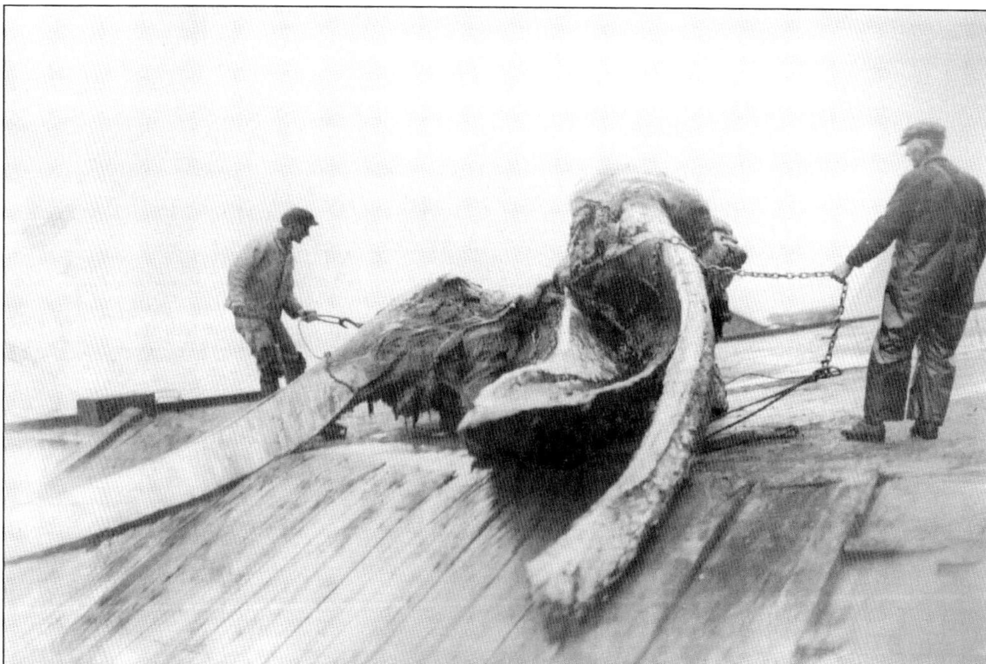

Jawbones and part of the head are being winched up the ramp to the meat loft and cookery for further processing. The bones are usually cut up and left outside until the end of the season or until they are ground to make fertilizer.

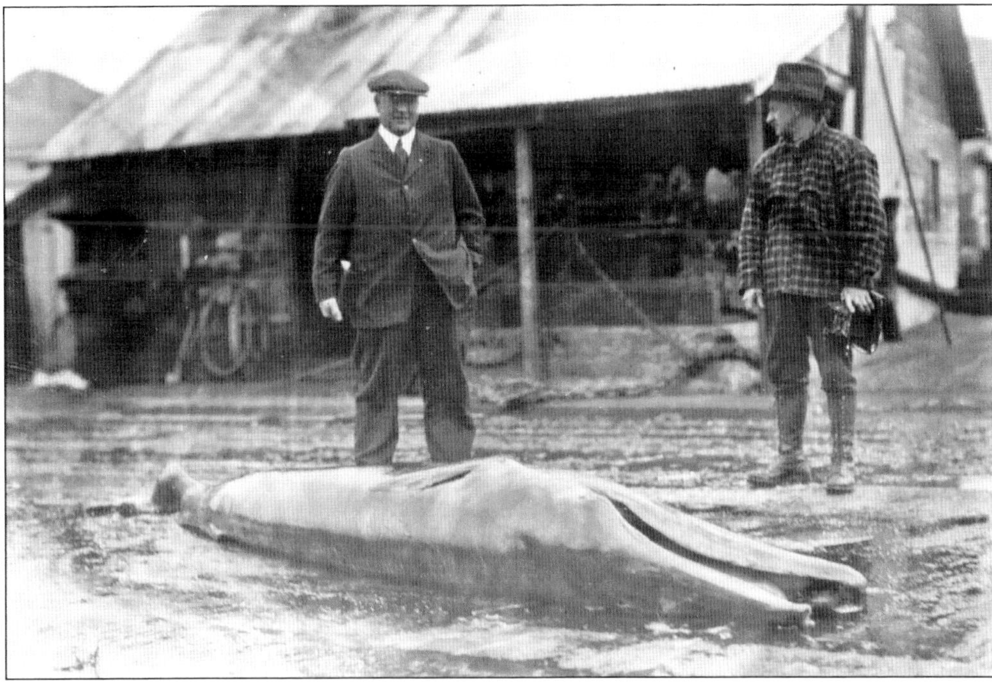

This is the fetus of a baleen whale that has been removed from its mother at the station. A fetus was normally processed like other whales. Regulations were enacted to prevent the taking of females with calves, but a pregnant whale many times could be taken.

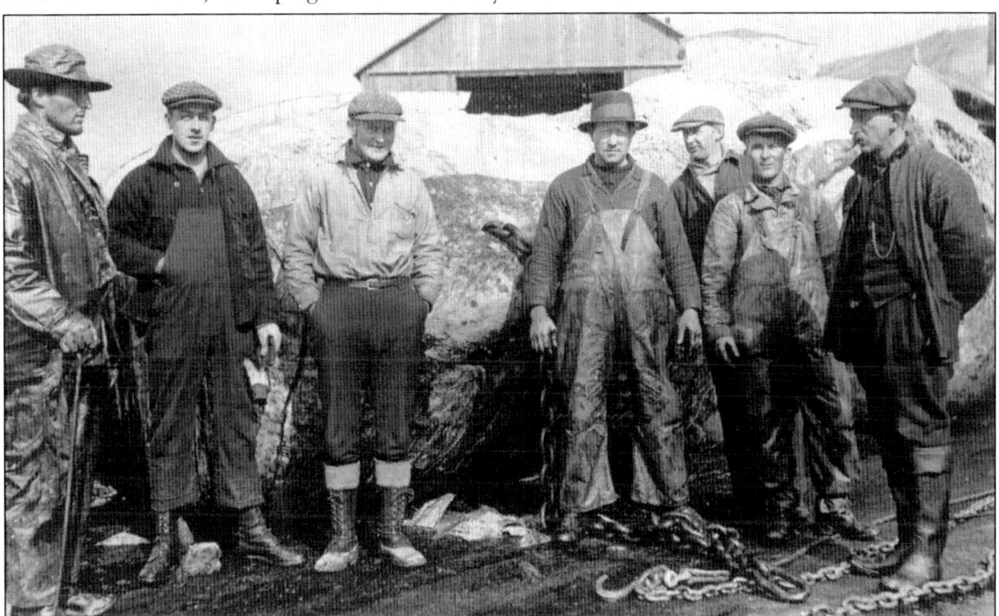

Seven men stand with a whale that has been flensed of its skin and blubber. The blubber is taken into the blubber factory, where it is reduced to oil and then put in 50-gallon drums for shipment to Seattle. In the early days at Akutan, the company would send its own ships or contract a ship to bring the products down. Later, regular service was available, and steamers would call at Akutan to pick up products for Seattle.

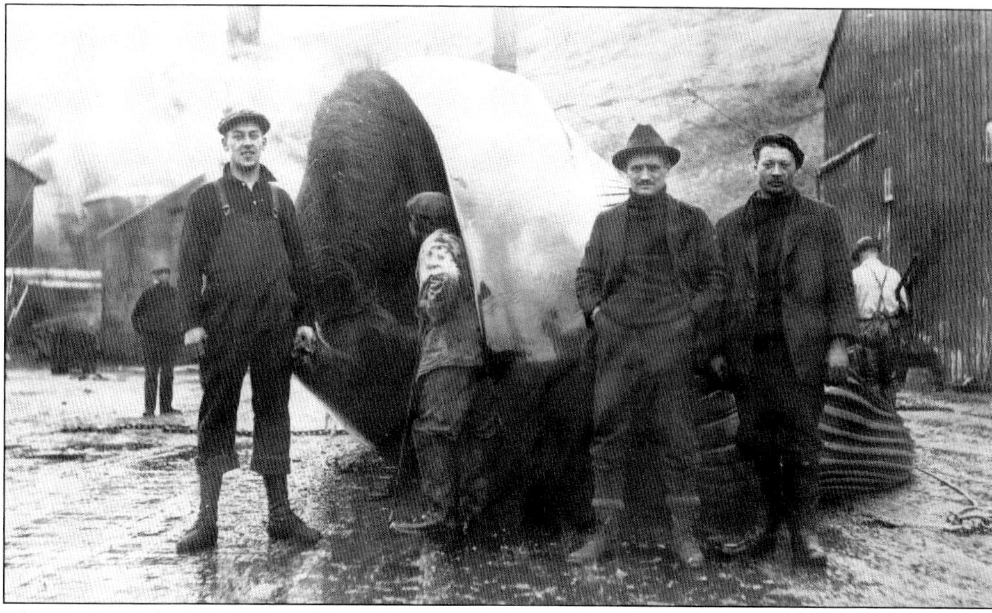

These men and a boy are standing with a baleen whale that has just been dragged to the blubber factory for removal of the baleen to be boiled down to oil. After this is done, the carcass will be towed to the next step in the process, which is to remove the meat. The entrails are removed before going on to the next step.

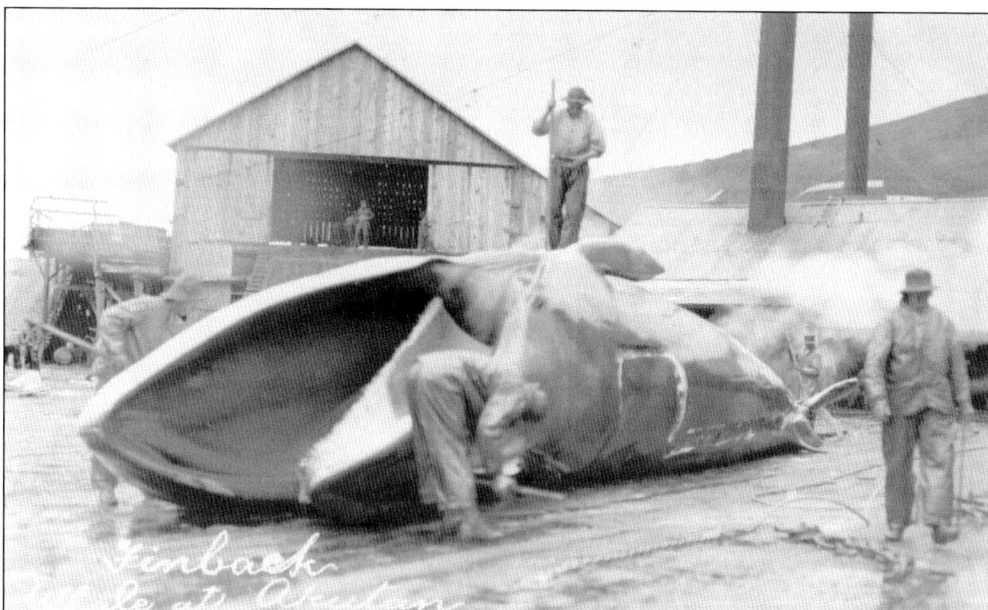

Flensers are making cuts in this finback whale carcass before the blanket blubber pieces are pulled off. In the back, up the ramp, is the meat loft, where the carcass will be taken after the blubber is removed. The building at right with the tall smokestack is the engine and boiler room. The winds in the area frequently blow the smoke and steam around the workers; this is visible in some of the Akutan photographs.

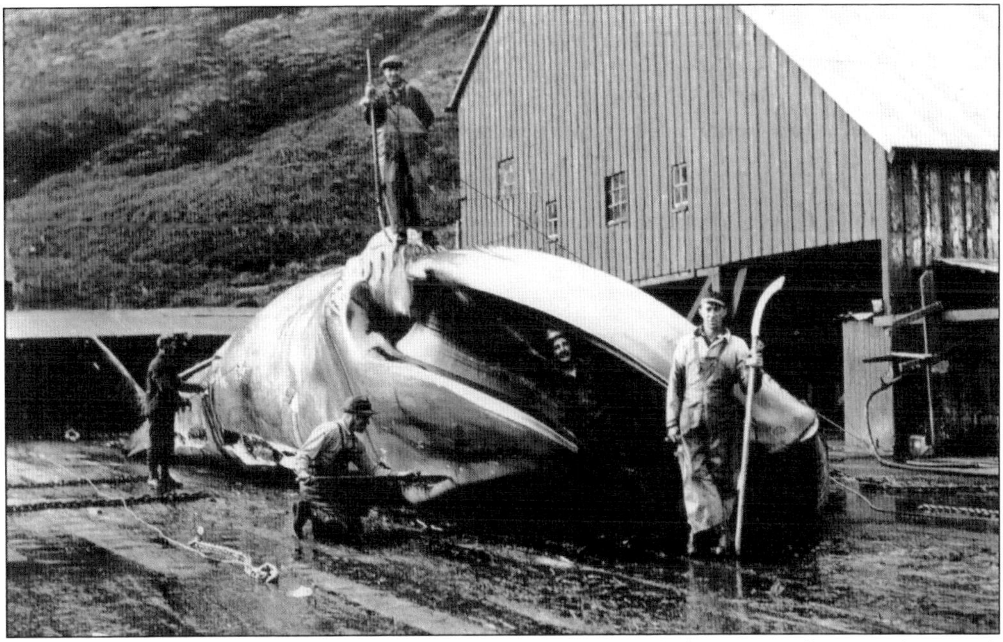

The team shown here is about to start in flensing this baleen whale. One worker is in the mouth of the whale with only his head showing. The building at right is where the blubber will be taken after it is removed from the whale.

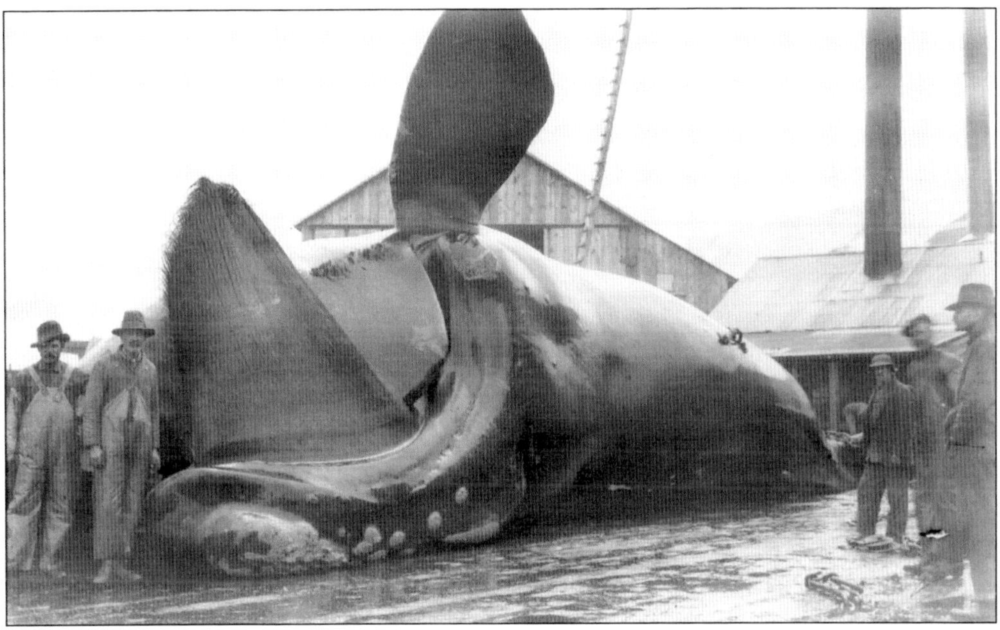

This bowhead whale has very long baleen, as well as white callosities on its head. Further down the body, a chain can be seen; this is where the harpoon entered the whale when it was shot. Sometimes, part of the harpoon may be sticking out of the whale, but it was not unusual for the harpoon to go all the way into the creature.

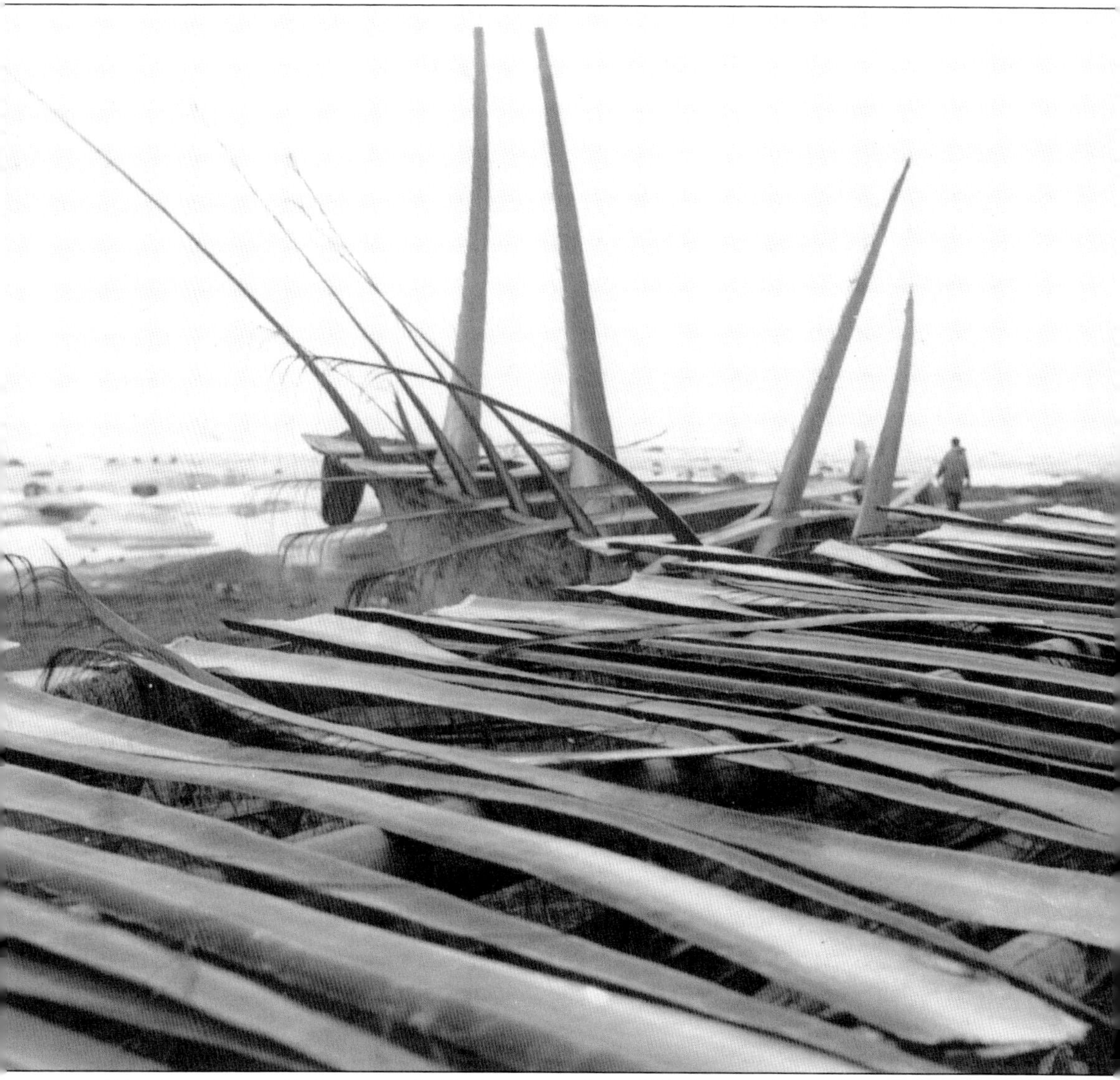

This is a picture of baleen—probably from one whale. Whales were hunted commercially for their oil and baleen, starting in the 11th century, and their oil was used for food products and lighting, while their baleen was used in corsets, parasols, and other products.

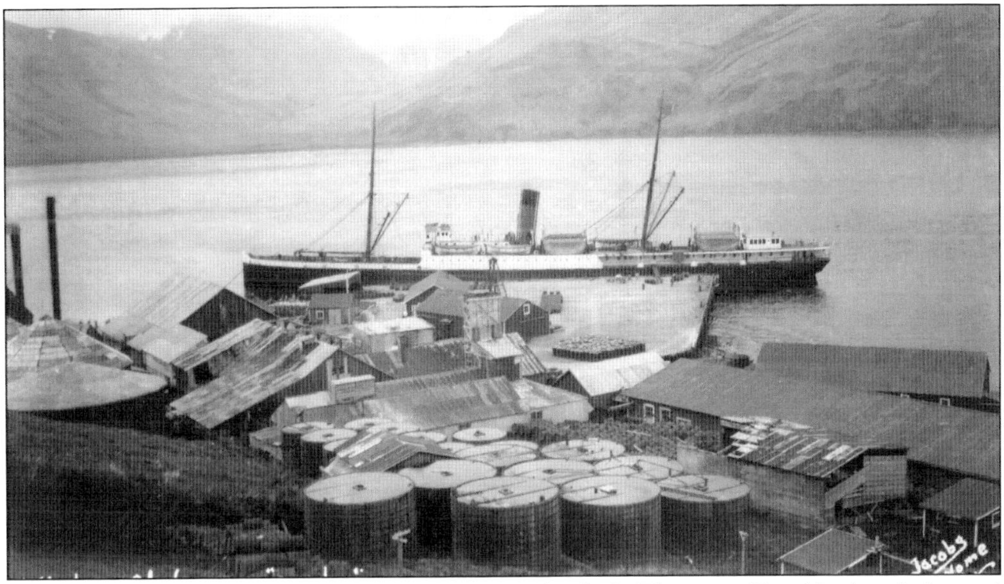

This station view shows an unidentified steamer at the dock. There are 17 or 18 oil tanks located at the rear of the station; next to them is a supply of barrels. The oil would be stored in the tanks and, after processing, put into the barrels for shipping. The tanks were later replaced with large-capacity tanks.

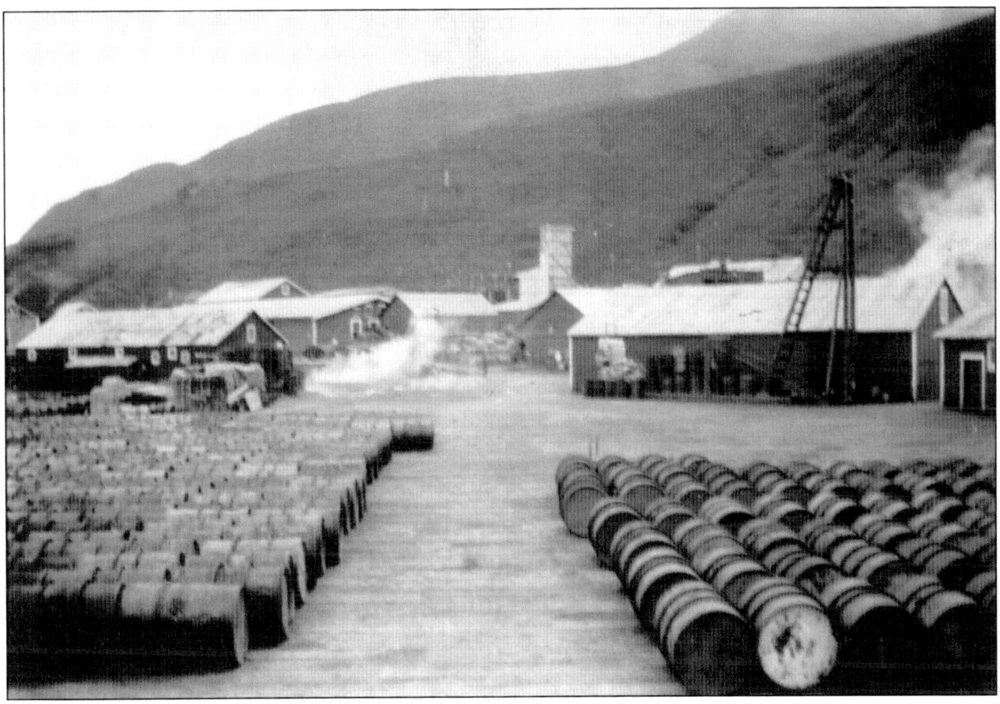

Barrels of various grades of whale oil wait at Akutan for shipment to Seattle. Until substitutes like vegetable oils and kerosene were developed, whale oil was widely used in lamps and to make soap and margarine. Sperm oil, which comes from the head cavities of sperm whales, is different from whale oils developed from blubber. It was used as a machine lubricant.

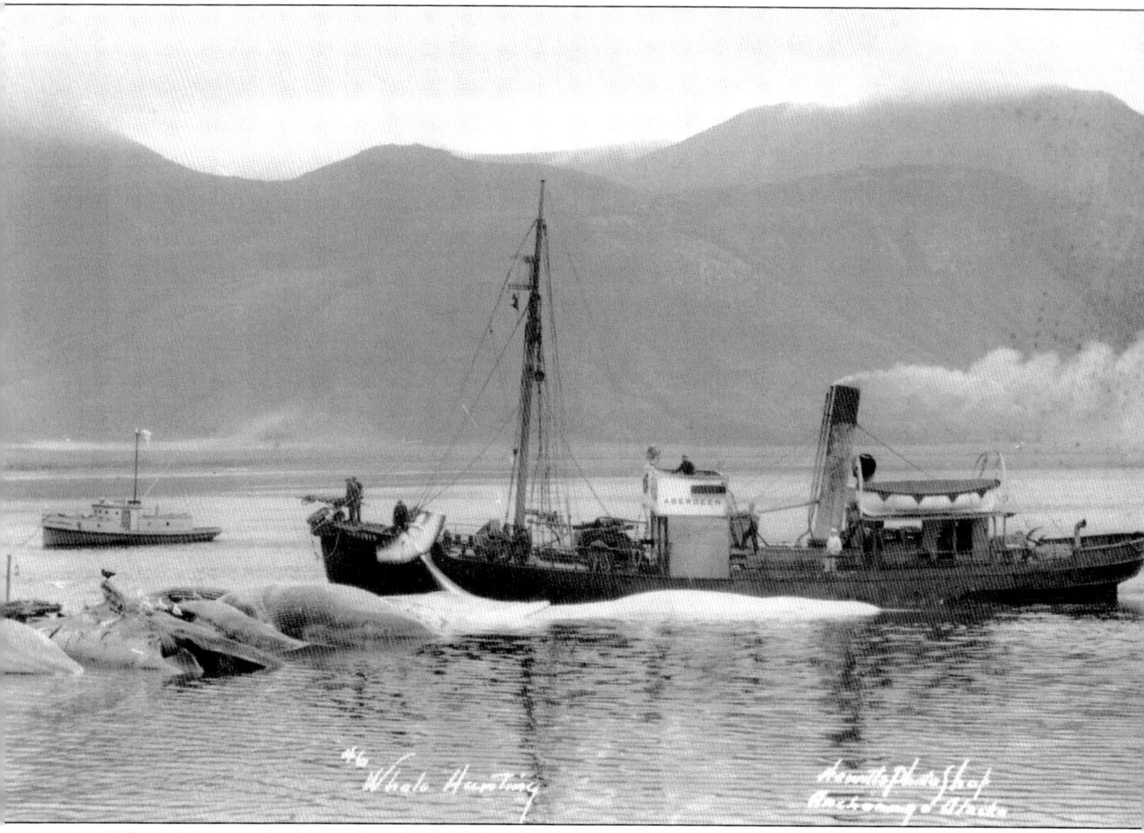

This photograph shows the whaler *Aberdeen* dropping off carcasses at Akutan. The *Aberdeen* was at Akutan in 1925 and 1926 and was then transferred to Port Hobron. In 1928, while returning to Seattle, it was caught in a gale that blew for six days. The captain put out the sea anchor and was in the storm for six days but had enough just fuel to make it to Juneau. After the storm and refueling, the *Aberdeen* continued on to Seattle.

Seven

Port Hobron Whaling

In 1925, the Port Hobron whaling station was being planned by the North Pacific Sea Products Company (NPSPC), the parent company of the Akutan station operating from the Aleutian Islands not far from Dutch Harbor. A major component of the plan was to use equipment from the Bay City station in Washington, which closed down after the 1925 season. A barge had been procured, and the reduction machinery from Bay City was loaded for the voyage to Alaska. Unfortunately, the tow boat encountered severe seas, and the cargo washed overboard. The equipment, however, was readily replaced with equipment from Kyuquot, Vancouver Island, and Grays Harbor stations, which also had recently been idled.

In late 1926, as reported the following year in *Pacific Fisherman* magazine, the parent company, "operating the Alaska plants . . . changed its name to American Pacific Whaling Company [APWC]." William Schupp, as the head of the new company, appointed his brother Henry to the post of station manager, where he remained until his death in the 1930s.

The new station at Port Hobron was constructed on Sitkalidak Island, 70 miles southwest of Kodiak. Henry Schupp, planned to have the plant ready for operation early in July 1926. Seven whaling vessels were in service at Akutan, and several of these were transferred to Port Hobron when the plant was ready. A new 5.5-by-50-foot Renneburg drier was purchased to replace the one that washed overboard. The plant was ready for operation around July 20 and experienced a very successful season. The new plant was hailed as the most up-to-date whaling station on the Pacific coast.

In 1926, the catch at Port Hobron was reported at about 243 whales, which was a very good record considering that the plant was not ready for operation until well into the season.

In 1927, Port Hobron took 258 whales, which yielded 455,600 gallons of oil. The *Westport*, working out of Port Hobron, was high boat of the fleet with a haul of 85 whales. The whalers were *Aberdeen*, *Moran*, and *Tanginak*. Olaf Nervick supervised operations at Port Hobron in 1927.

Leaving Port Hobron on October 13, 1928, the small whaler *Aberdeen* ran into a terrible storm in the Gulf of Alaska. Capt. A. Halfred was forced to heave-to and ride at sea anchor for six days in a gale that mounted to Force 12, with the barometer dropping below 28.90. With his fuel oil nearly exhausted, the skipper was forced to drive his ship into head seas in an effort to reach Juneau under steam. In spite of grave danger of foundering, the *Aberdeen* wore through and reached Juneau with a narrow margin of fuel to spare. In 1928, the APWC had a fair season in northern waters. The Port Hobron station, with 256 whales, was again the lead station.

The seven killer boats of the APWC sailed from Seattle on May 15, 1929, for western Alaska waters. The vessels were manned with 10 men each, *Pacific Fisherman* reported, "while the personnel for the company's shore stations at Akutan and Port Hobron followed on passenger steamers." The Port Hobron fleet was made up of *Tanginak* (Captain Murray), *Aberdeen* (Captain Peterson), and *Moran* (Captain Aaset).

In 1929, the APWC operated stations at Akutan and Port Hobron with good results. Whale hunting in the north was very successful early in the season, and by early fall, the catch was well ahead of what it had been the previous year at the same time.

In late May 1930, *Pacific Fisherman* reported that an 86-foot blue whale was struck by the killer boat *Tanginak* of the APWC in Alaska waters and "gave trouble reminiscent of the old days of hand whaling and occupied the killing steamer for an entire day. The first harpoon failed to strike a vital part, but held. Another shot was fired into the whale, and this also remained fast without killing the creature, which towed the steamer all day before the whale was exhausted and the [*Tanginak*] could be maneuvered into position to place the fatal shot."

All of the whaling stations remained closed for the 1931 season. The APWC closed its stations at both Port Hobron and Akutan. The suspension of whaling was due to the inability of the whaling companies to secure contracts for whale products. Port Hobron, however, reopened in 1932, with Marc A. Lagen as the station manager. He was the son-in-law of William Schupp and took over from Henry Schupp, the previous manager, who died while the plants were shut down.

In 1932, the Port Hobron station of the APWC was operational on Kodiak Island. It was the first whaling operation in the North Pacific in two years. According to the *Pacific Fisherman*, four killer boats went north from Seattle in May to serve the station: *Aberdeen* (Capt. John Berntzen, master and gunner), *Westport* (Capt. A.M. Storvick, master, and Chris Olsen, gunner), *Paterson* (Capt. Peder Oness, master, and T.S. Christianson, gunner), and *Moran* (Capt. H. Aaset, master and gunner).

The catch of whales for 1932 at Port Hobron was 269, and the station produced 603,883 gallons of whale oil and 5,780 gallons of sperm oil. The catch for 1933 was 237 whales.

In 1934, for the first time since 1930, the APWC ran its plants at both Port Hobron and Akutan. Lagen was in charge of operations at Port Hobron, with three killers: *Aberdeen* (Capt. J. Berntzen, master, and Gus Wester, gunner), *Moran* (Capt. H. Aaset, master and gunner), and *Tanginak* (Capt. A. Darron, and Chris Olsen, gunner). Port Hobron took 137 whales—exactly 100 short of the 1933 catch. In 1935, Port Hobron increased its take from 137 to 188.

In 1936, Capt. Albert Barron commanded *Tanginak* and Capt. John Hartland *Aberdeen*, with Gus Wester as gunner. For the season, the station took 120 whales, with the *Moran* taking 32 whales, the *Aberdeen* taking 51 whales, and the *Tanginak* taking 37 whales.

Port Hobron whaling station did not operate after 1937. During World War II, much like in World War I, some of the whaling vessels were taken for military service in the area.

William Schupp died in 1948 at the age of 82. The Akutan and Port Hobron properties were sold for a total of $500. The vessels, *Aberdeen*, *Kodiak*, *Moran*, *Paterson*, and *Tanginak*, were sold for a total of $4,550. The company announced that it would never engage in whaling operations again.

In a brochure of the Baranov Museum on Kodiak Island, Anjuli Grantham, curator of the museum, writes, "During the 1920s, the *Carolyn Frances*, the Erskine family charter vessel, whaled around Kodiak. Whales were butchered alongside the boat and the blubber was rendered into oil on deck." The Erskine family was friends with Capt. Louis Lane, who would take the family whaling when he was around Kodiak Island. Captain Lane also sold some of the whale carcasses to local fox farms in the vicinity of Kodiak Island for only $500 each.

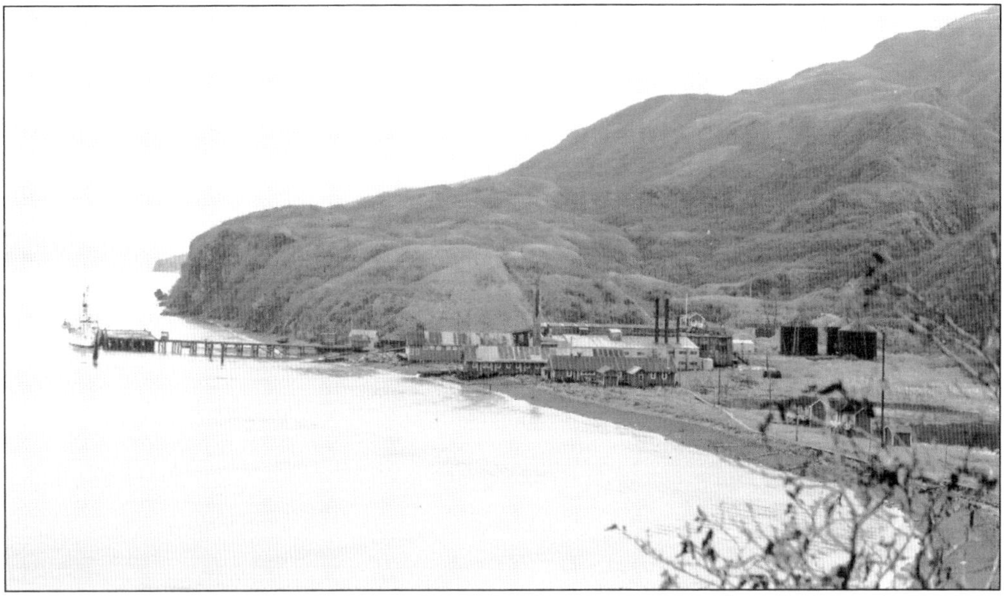

Here is a photograph showing the Port Hobron whaling station, with a whaler at the dock on the left, the factory in the center, and the oil storage tanks on the right. The station was on Sitkalidak Island, 70 miles southwest of the town of Kodiak, Alaska. (Courtesy of Kodiak Historical Society, P552-6.)

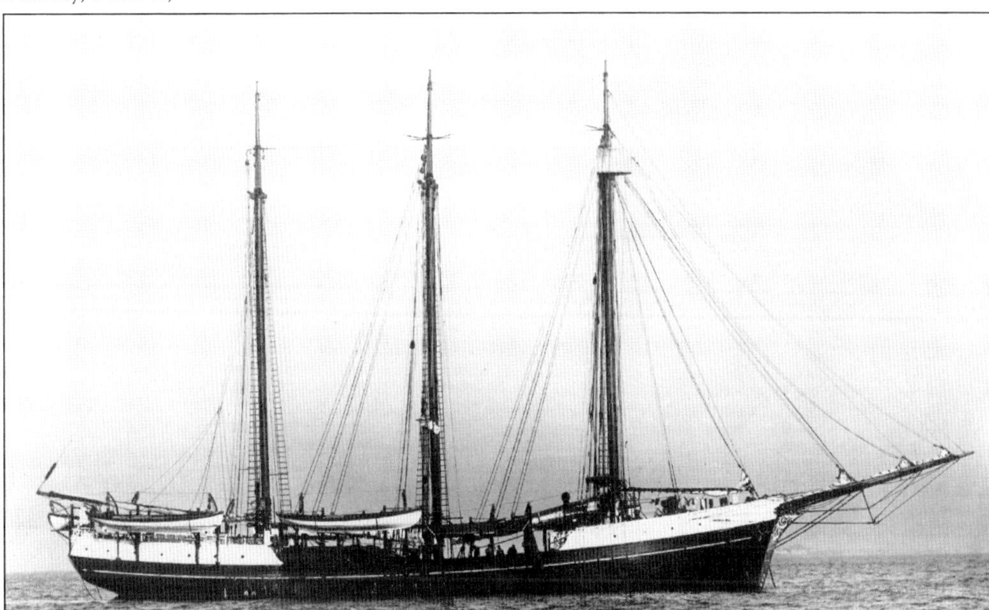

Carolyn Frances sits in the stream off San Francisco. It was owned by the Alaska Commercial Company, was chartered out of Kodiak, and went whaling and trading on the coast of Alaska. It was named for the daughter of the owner of the Alaska Commercial Company, Louis Erskine. It was rented out by the Northern Commercial Company, with a home office in Kodiak, to film a movie about whaling around the area. Louis Lane also rented the *Carolyn Frances* from the Northern Commercial Company in the 1920s to use for whaling and trading. (Courtesy of San Francisco National Historic Park.)

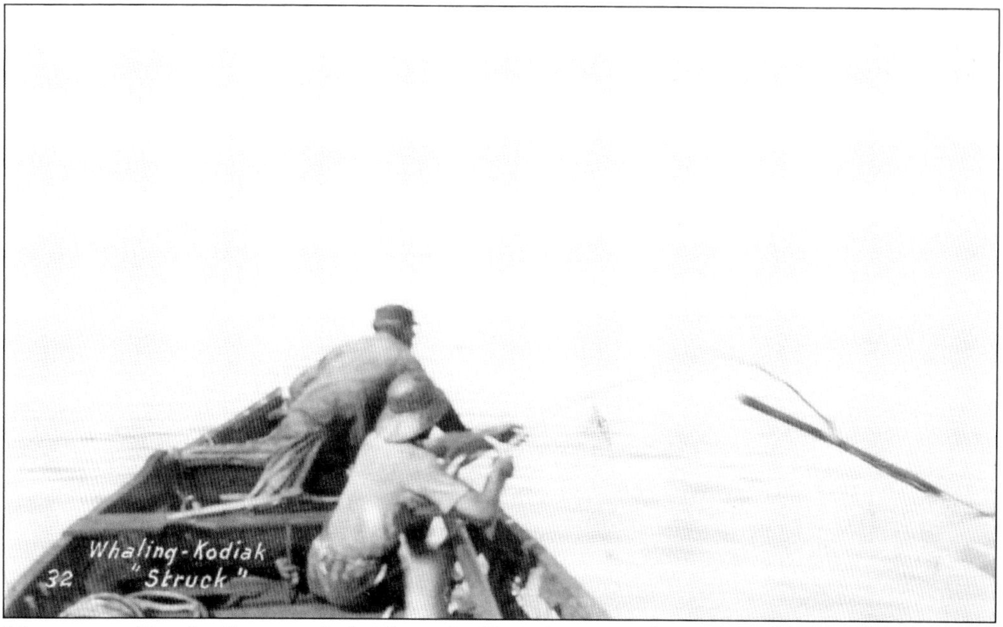

The steamer *Carolyn Frances* whaled from the Kodiak area in the 1920s. This photograph shows a harpoon being hand-thrown at a whale from a wooden whaleboat. (Courtesy of San Francisco National Historic Park.)

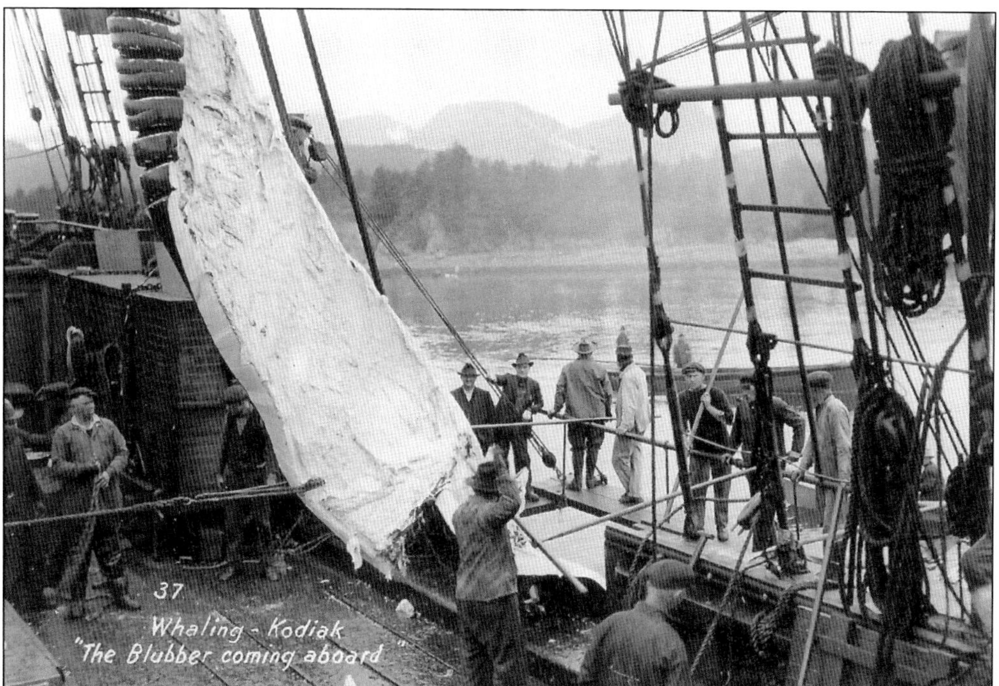

On the *Carolyn Frances*, these men are lifting a blanket piece that was cut off around the whale as opposed to what usually occurred at a shore station, where workers cut the blanket piece from the head of the whale to the tail. (Courtesy of San Francisco National Historic Park.)

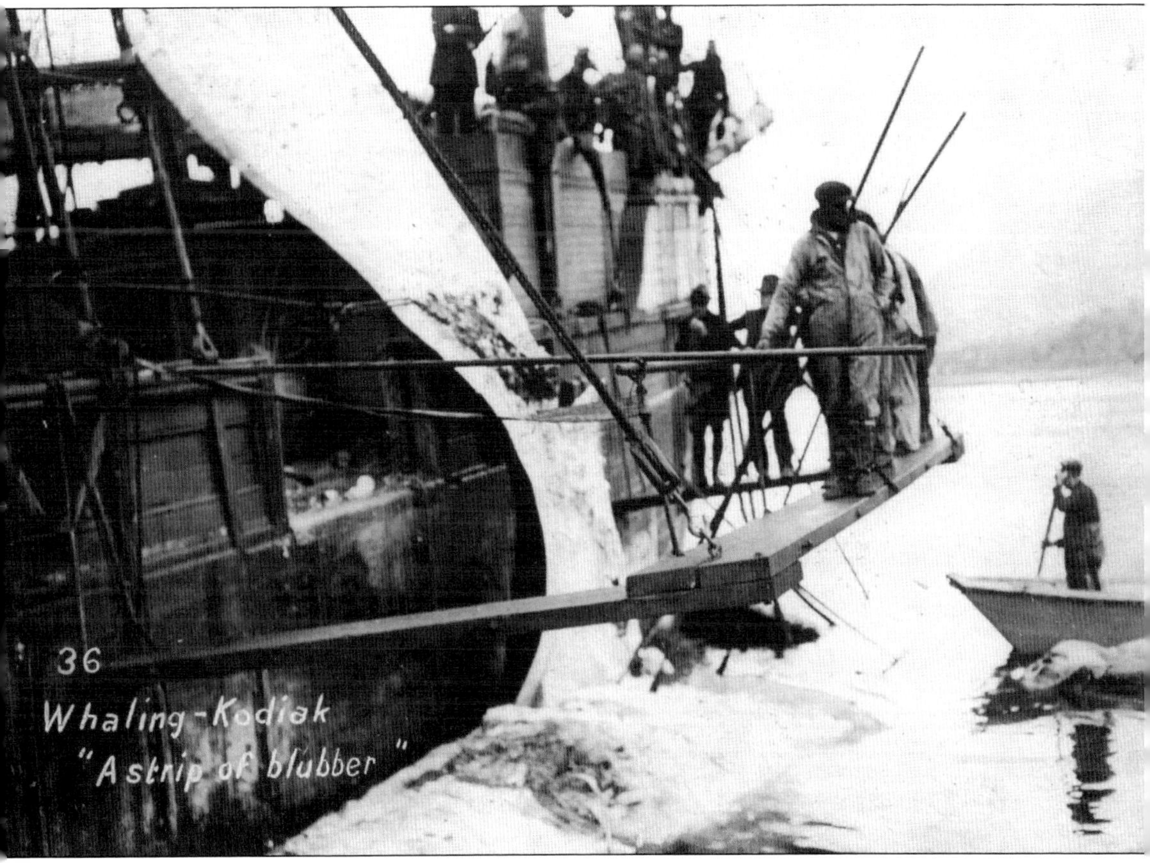

Captained by Louis Lane, the *Carolyn Frances* was a modern incarnation of the offshore whaling vessel from centuries past—whales were flensed alongside the boat instead of being taken back to shore for cleaning and cutting.

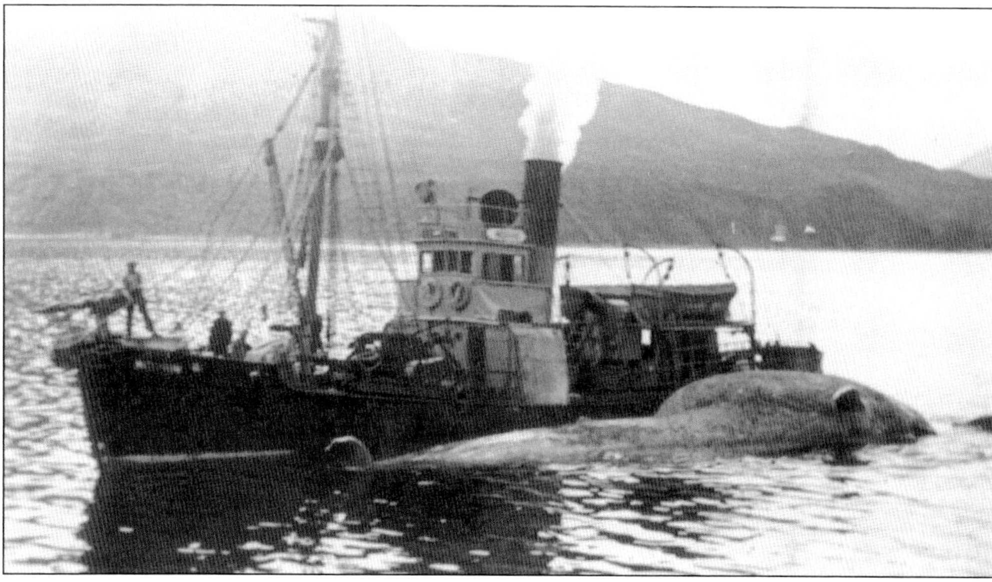

The whaler *Moran* arrives at Port Hobron whaling station with a whale in tow. The whale will be tied up at the station or taken directly to the station's ramp. The *Moran*, which was 96 feet long, was built in Seattle in 1911 by Seattle Construction and Dry Dock Company.

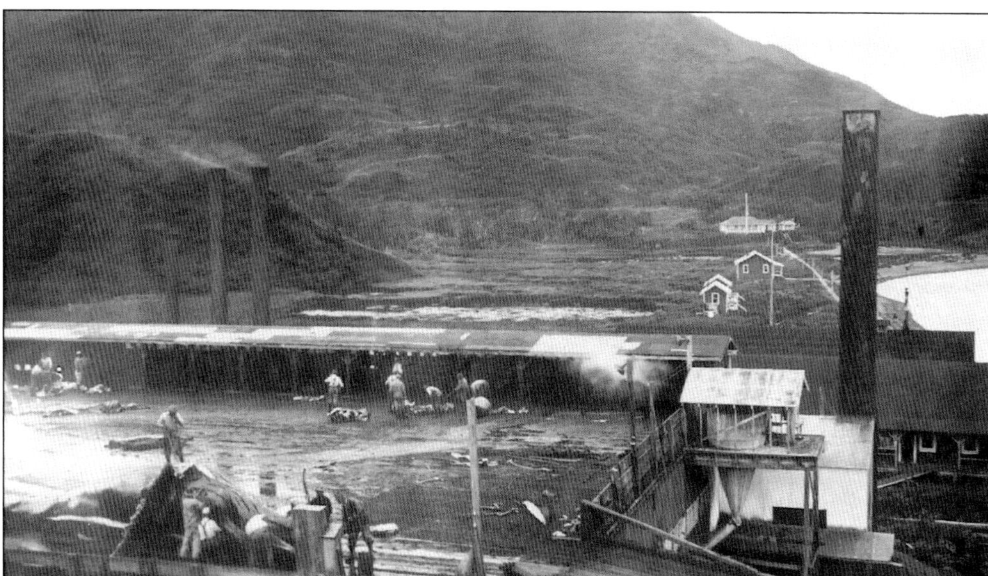

This view of the station shows the mostly open-air flensing area, where the bones, meat, and blubber are separated from each other for further processing. The whale on the platform (at lower left) is probably a baleen whale.

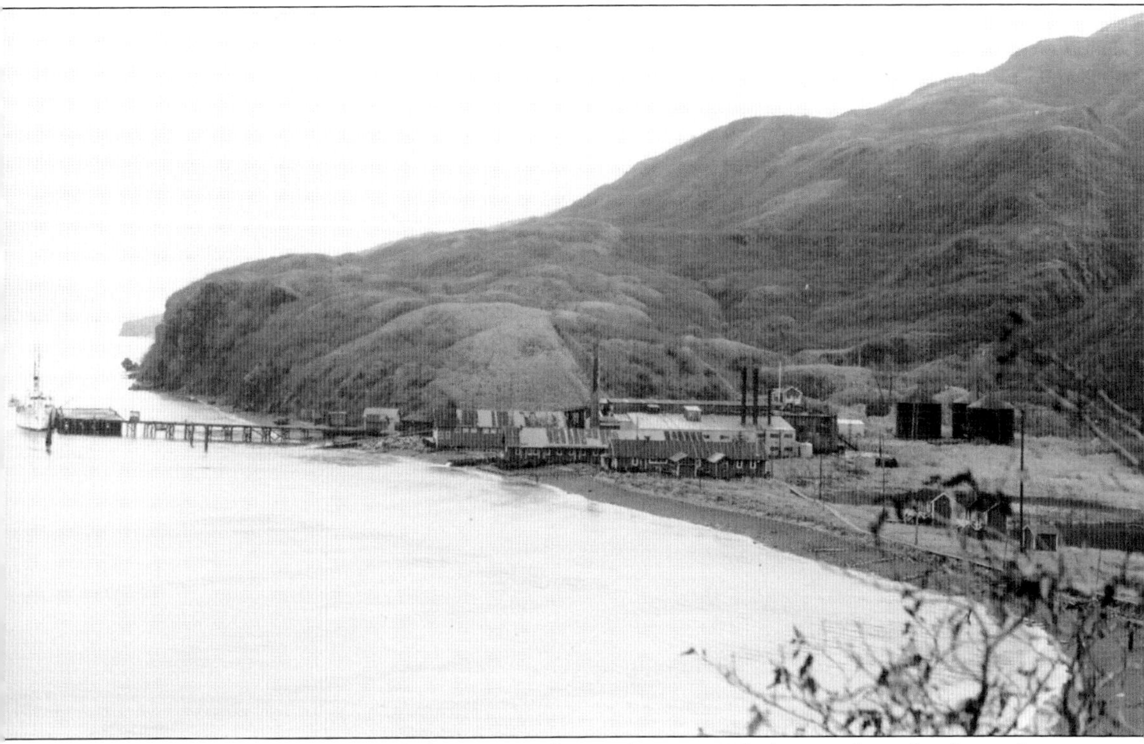

This is a view of the Port Hobron whaling station from a distance. Whales were brought alongside the dock and to the ramp, then moved from left to right through the station during processing. Oil ended up in the three tanks shown at right. The main part of the station is in the center of the picture, but some time after the station closed, it caught fire and burned to the ground. The three oil storage tanks are still standing.

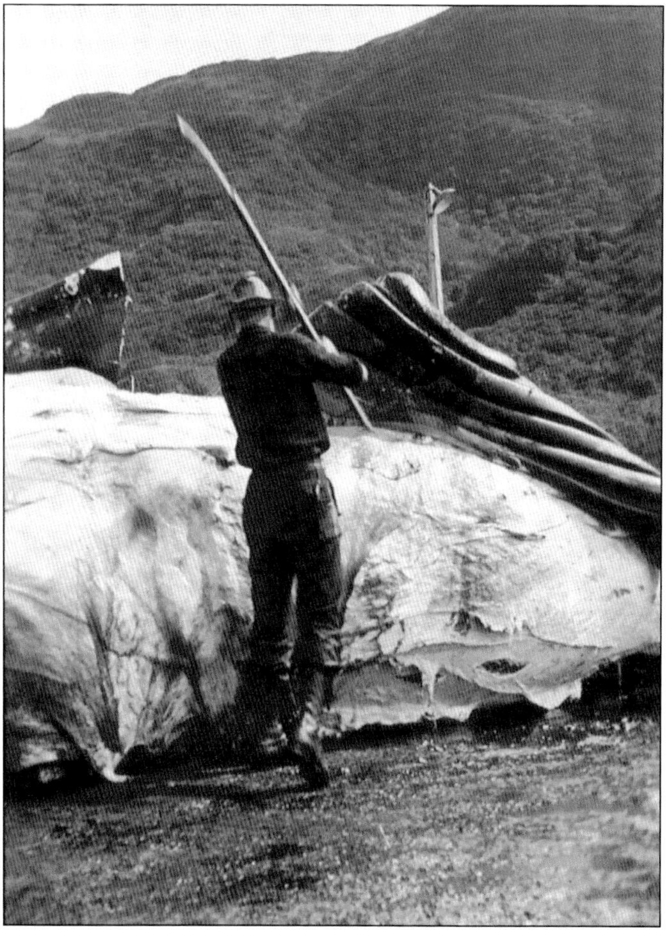

This man is working to remove blubber from the meat and bones. The blubber will be taken to the cooker to try out the oil. In the 1800s, a whaler's treat involved some bread dough put in the cooker to create the whaler's interpretation of a donut. The residue left when all the oil is boiled out is basically what most people know as cracklings. Some people might snack on the cracklings, but most of these remains were thrown into the fire.

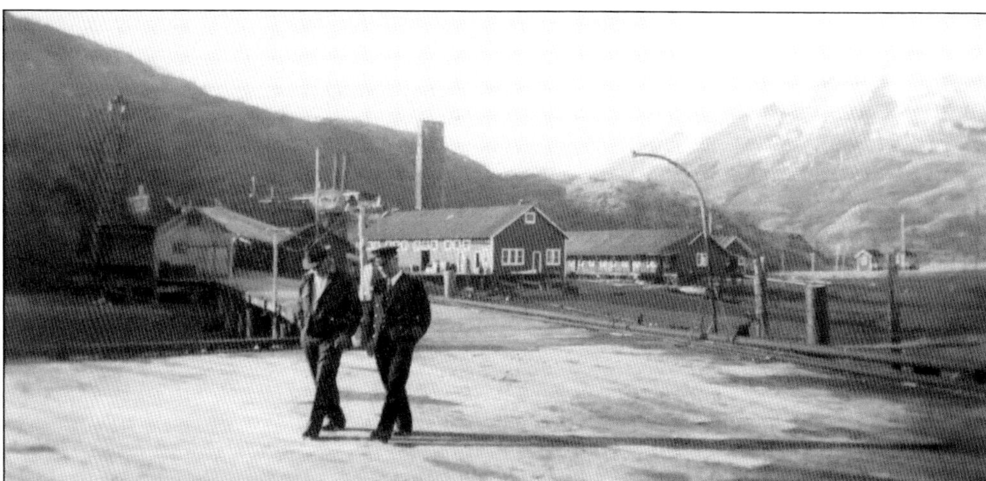

This view of the Port Hobron station shows the residence area. The station operated from 1926 until 1939 but was idled during 1931 and from 1937 to 1939. After 1939, neither of the Alaska stations—Akutan or Port Hobron—were operational. The station catch ranged from about 120 to almost 300 whales per season.

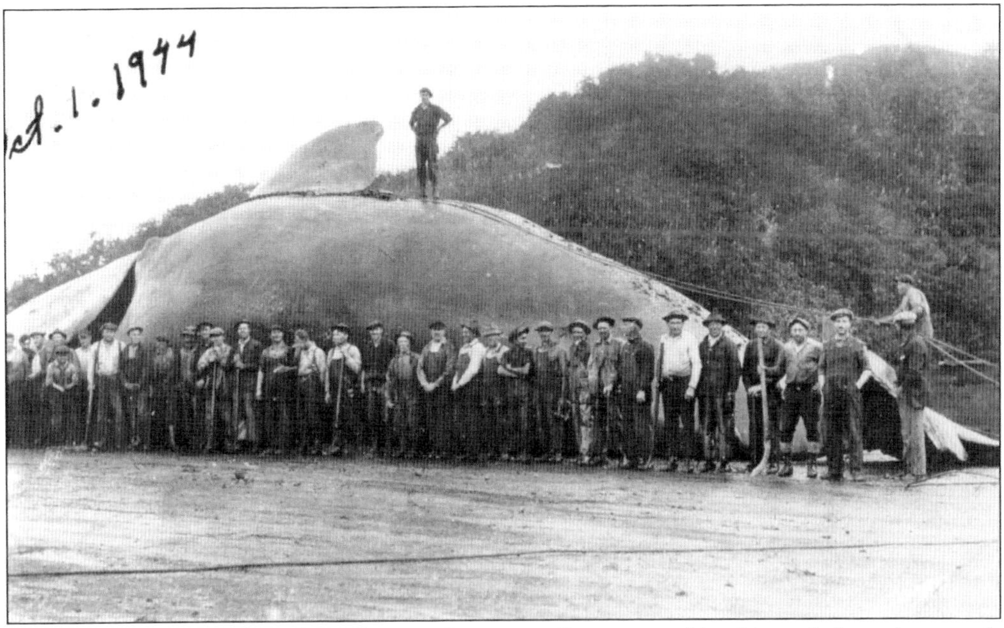

Here, a right whale has been brought in to the station. Rare events like this made for a photograph opportunity. All of the workers turned out for the event, with the head flenser standing on top of the whale.

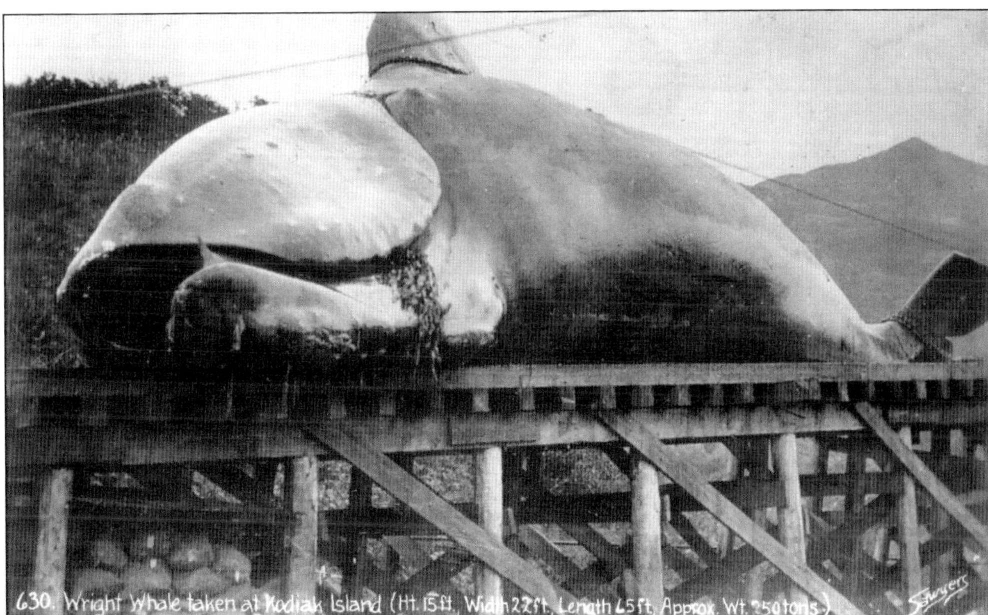

Right whales, like this one, can grow up to 59 feet long and weigh up to 100 tons—significantly larger than humpbacks or gray whales, but smaller than finback and blue whales.

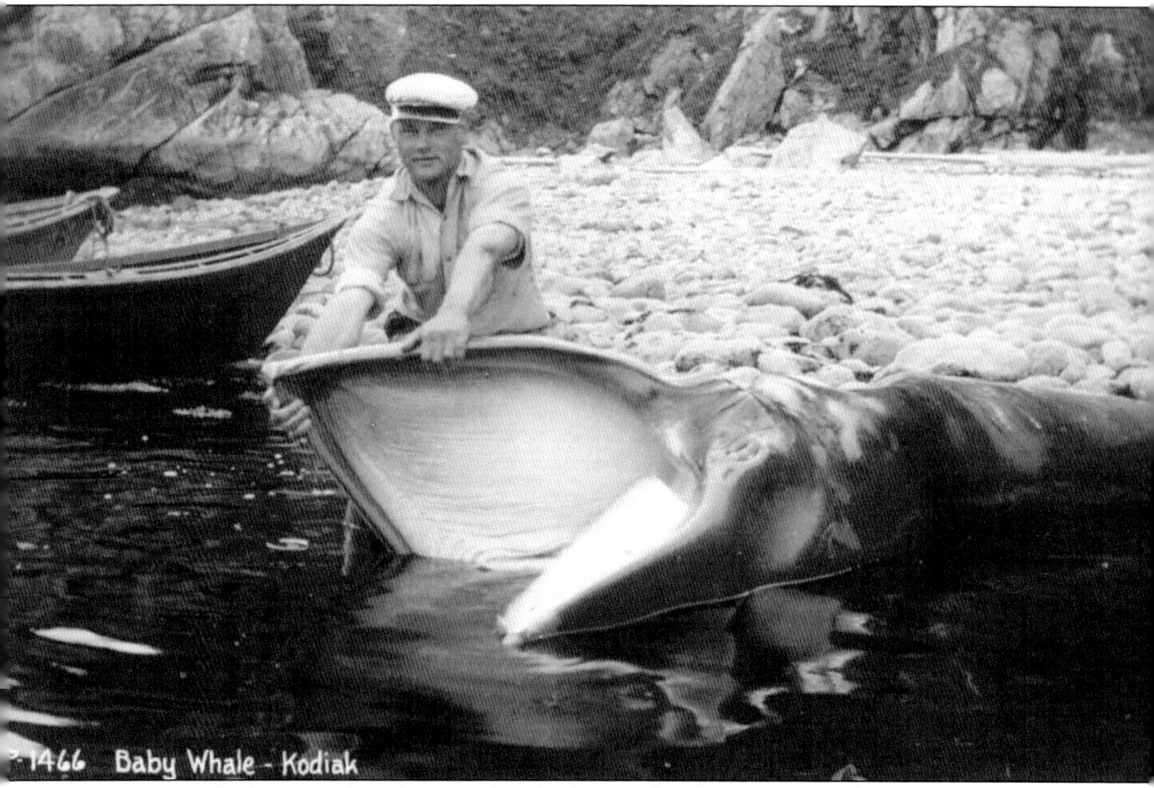

This small whale was found on the beach; the unidentified man is holding its mouth open to show the baleen plates. Instead of teeth, baleen whales have baleen plates used for filtering food from water. The humpback whale is known for its acrobatics, but other baleen whales also break through the water surface with their bodies or beat it loudly with their fins.

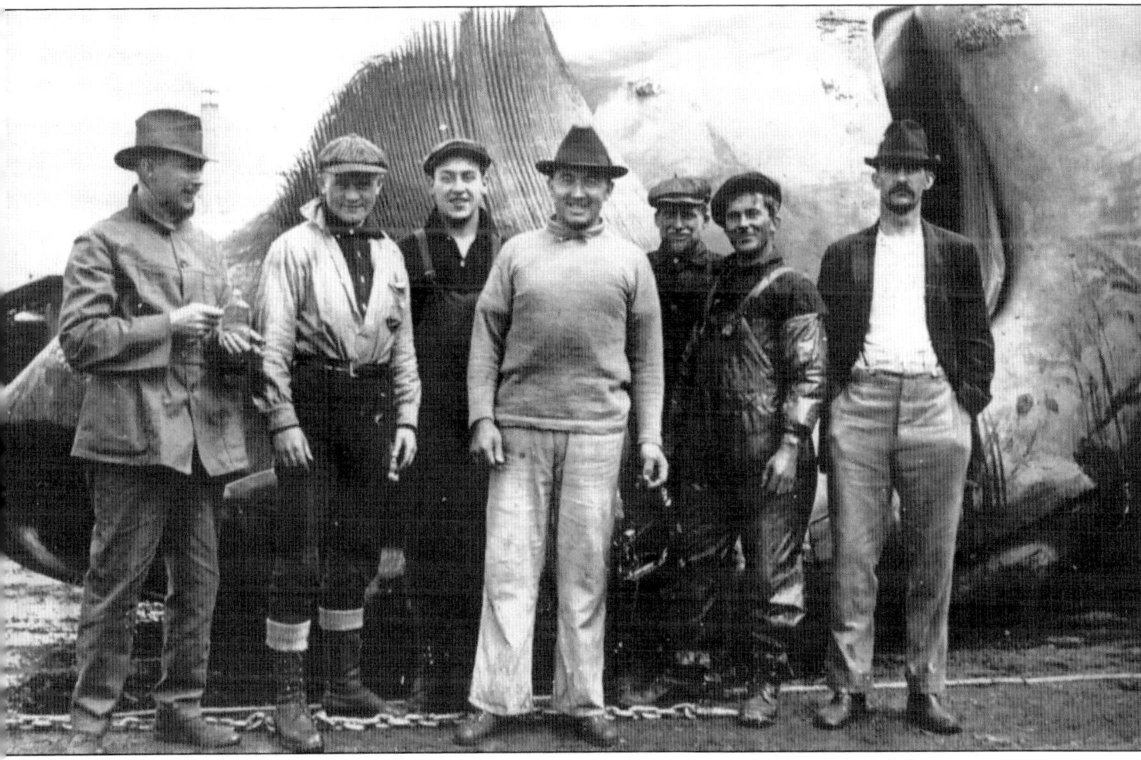

This group of men poses with a bowhead whale. The baleen plates, shown behind the men, can be as long as 15 feet.

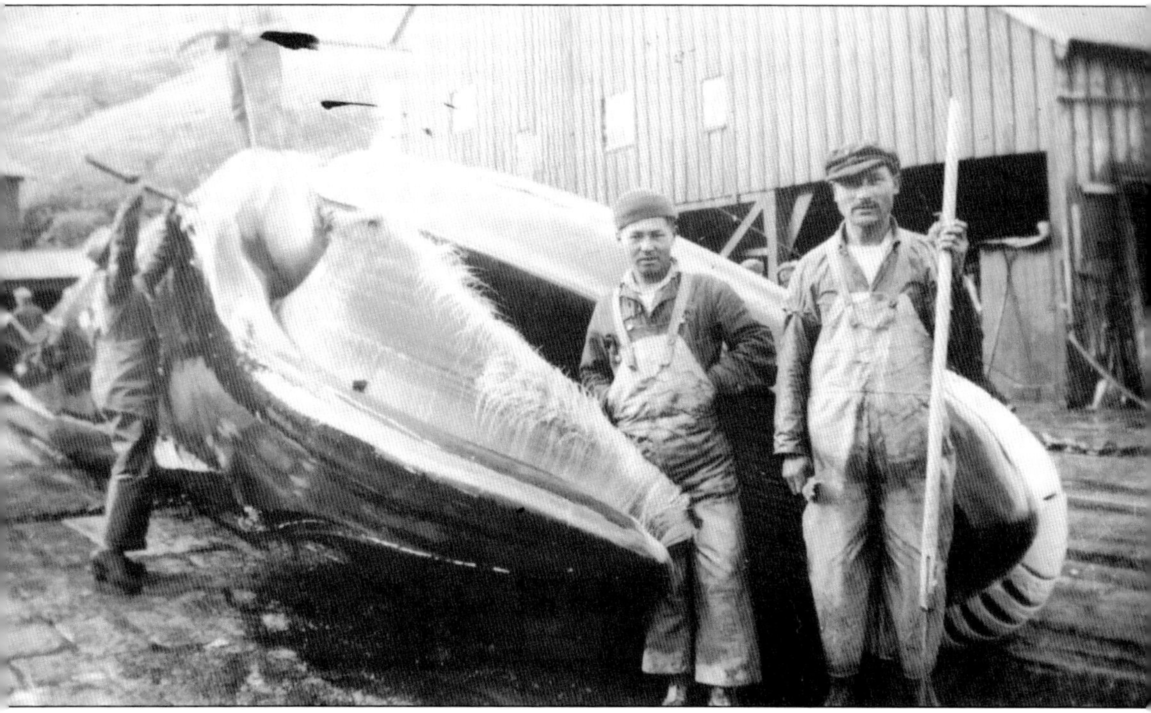

These workers are standing with what appears to be a finback whale. The finback is the most streamlined of the whales, with a black lip on one side and a white lip on the other. The baleen in baleen whales continues to grow and frays on the end as it is used. The frayed ends result in a mass of fibers that help to trap small krill and other food in the whale's mouth.

Eight

Modern Shore Whaling

The *Report of the Commissioner of Fisheries* for 1909 provides a good description of the whales, catching methods, and processing is provided. This process changed very little from the earliest days in Alaska until the close of whaling with the start of World War II.

> Three species of whales are sought by the whalers from the shore stations operating along the North Pacific coast . . . the sulphur (blue whale), finback, and humpback. During May to September, these whales are often found in large numbers close to shore. They yield a large quantity of oil, and 800 pounds of baleen, 3 feet long. . . . The finback, or finner, approaches the sulphur bottom whale in length, in some cases measuring 70 feet, without as much bulk. The finback is acknowledged to be the fastest whale that swims. Its back is colored a blue-black, turning to almost white underneath. The flippers are comparatively short, the dorsal fin prominent and situated nearer mid-way of the animal's back than in the other baleen whales. The baleen, about 2 feet 6 inches, is colored a light bluish gray, streaked with black. A fair quantity of oil is secured from this whale. The finbacks are very erratic in their movements and the fishery consequently uncertain.
>
> The humpback, the most common species caught by the whalers, is one of the whales that roam through every ocean, generally preferring to feed and frolic nearer the coast. Short and of ungainly bulk, the humpback is equipped with flippers and flukes seemingly out of all proportion to its body, the flukes often measuring 25 feet across. It's under jaw extends forward considerably beyond the upper one. The top of its head is dotted with irregular, rounded bunches. The color is black above and white underneath. Marine parasites in the form of very large barnacles are always found upon it in numbers.

Whaling steamers were designed to be quiet, moving through the water with minimal friction so as not to announce their arrival to the whales. At the bow was a heavily constructed Svend Foyn harpoon gun with a three-inch bore. The harpoon weighed several hundred pounds, and screwed into it was a bomb containing about a pound of gunpowder. The "foregoer" was coiled under the muzzle, with one end attached to the harpoon and the other to one of the main whale lines, which may reach up to 400 fathoms in length and be over five inches in circumference.

The struggle to secure the massive animal is described next:

> As soon as struck, the whale sounds and goes to the bottom. The animal has immense strength and will at times tow the steamer several miles before beginning to weaken. As soon as the line slackens it is snubbed around a heavy steam winch on the deck just ahead of the bridge, after which the wounded whale is played by the expert angler, a continual strain being kept on him, slacking sometimes to avoid a wild rush, but always reeling in slack at every opportunity. The strain soon begins to tell on him, his rushes growing shorter and less vicious, and finally he rises to the surface, lashing the water in his struggle.

> The line is now rapidly hove in until the slack and the weight of the whale is showing, then the line is run through a heavy iron block at the foremast head, this mast being heavily rigged in order to stand the tremendous strain. Fathom by fathom the line comes in until the dead body is alongside. A chain is attached around the tail and the winch then heaves the tail out of the water, causing the animal to hang vertically head downward from the bow. The steamer is then forced ahead at full speed, to bring the body to the surface. The lobes of the tail are then severed and brought on board. In order to make the carcass more buoyant air is blown into the abdominal cavity.

Having arrived at the station, the whale was hauled to the mouth of the slip by way of a line connected to a buoy. The slip itself was sandwiched between two rock-filled cribs. A hefty iron chain, fully one inch and a half in diameter and powered by steam winch, drags the whale by its tail out of the water and to a location under the flensing shed.

Cobb and Kutchin's *Fisheries of Alaska* (1908) gives a detailed description of what happens to the whale next:

> As soon as the whale is in place men with long-handled knives commence "flensing." That is, removing the blubber. The men walk from the head toward the tail cutting long gashes in the blubber as they go, then a steel hook attached to a wire cable is hooked in at the end of a strip, the steam winch heaves on the wire, and the long strips are peeled off one after another.
>
> These strips of blubber are put into the slicer, or blubber cutter, and chopped into half-inch slices, which are hoisted to the blubber pots, where the oil is tried out by means of steam pipes running through the pots. After the blubber is exhausted in these pots, it is conveyed in a chute to a drainage tank, where the bulk of the water is separated by gravity, and then to the dryer, where it is mixed with the residue of the meat, and is turned into guano.
>
> After the blubber is removed from the carcass and the inside fat is taken out, the carcass is hauled up to the carcass platform. Here another gang of men remove the meat from the skeleton. This meat is put into pots, where it is boiled and the oil extracted from it by acid processes. After the oil has been dipped off from these meat pots, it is run into the drainage tank, then going into the hot-air dryer with the blubber residue. Here it is made into guano by a drying process which dries the material thoroughly and then shreds it fine, after which it is ready for the market, its value as a fertilizer being very high.

Once the blubber oil cooled, it could be barreled. The rest, however, must first have all meat remnants removed, even the smallest particles, in a processed called clarification. The various whale parts are made into a number of products—tails, sliced and salted, were sent to be consumed as jerky in Japan, for example, and glue was made from residual blubber that the stations would use for coating the insides of oil barrels. And the skin, of course, made excellent leather.

"If the prejudice against whale meat could be overcome it could have proved a most important addition to the larder, in Alaska at least, where fresh meat is difficult to obtain." Cobb and Kutchin conclude, saying that choice cuts can be found in the tail and beneath the eye. "It is said to have much the flavor and appearance of beef."

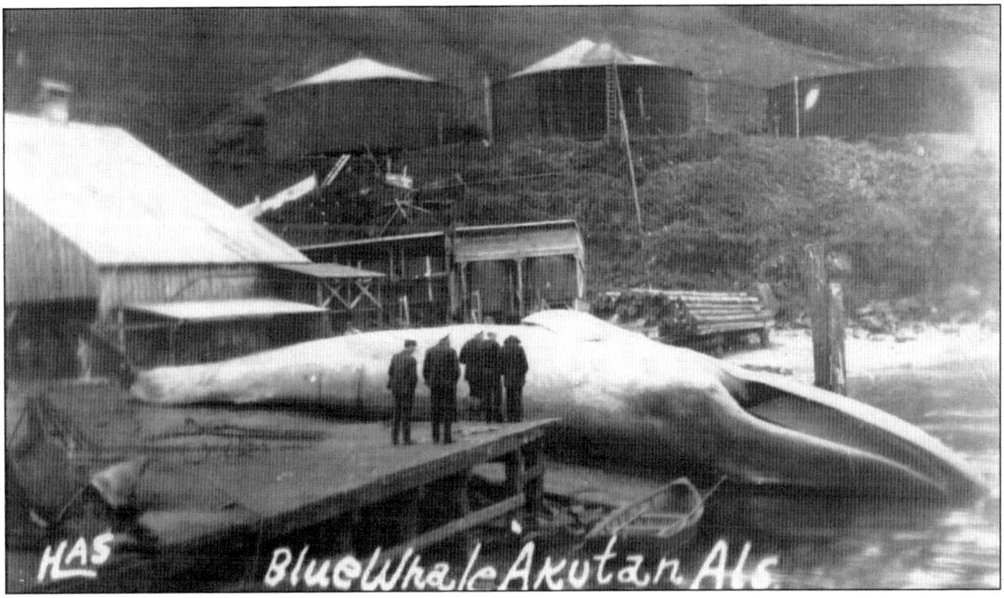

These men are inspecting a large blue whale that is at least 80 feet long. The blue whale is not only the largest whale found on the coast but also the largest known mammal, with the length of an adult varying from 60 to 100 feet. The origin of the whale's early "sulphur bottom" name is not clear, with some sources stating that it was derived from a yellowish cast to the skin on the lower side of the animal.

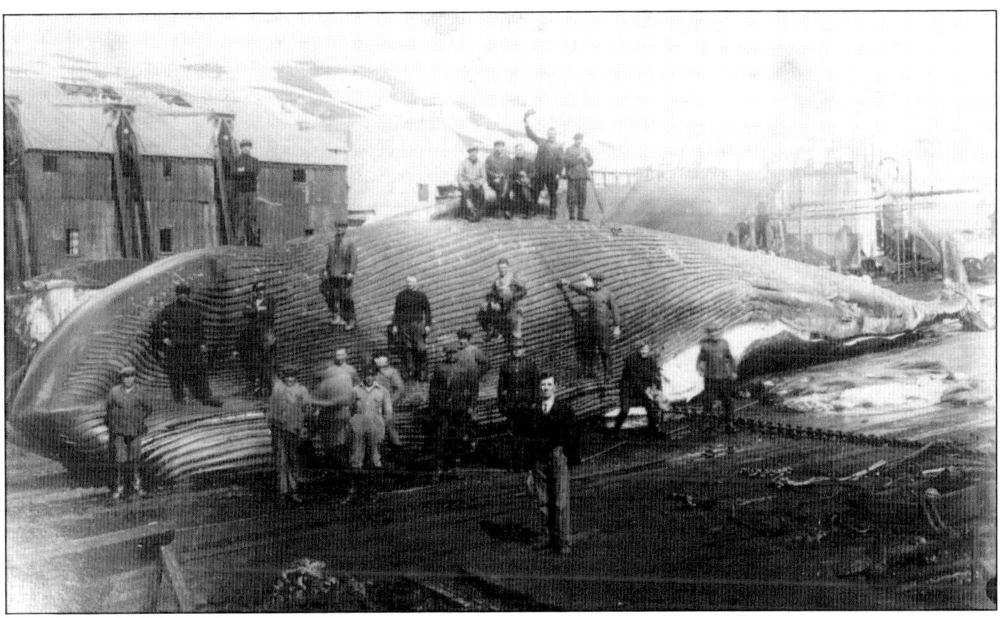
Employees of the whaling station pose with a large blue whale. Note that the head flenser is in his usual position on top of the whale, holding his flensing knife. The station manager is usually in front in the middle or off to the side, but they wear ties. The employees came up from San Francisco on the flensing boat or steamers after regular traffic was established.

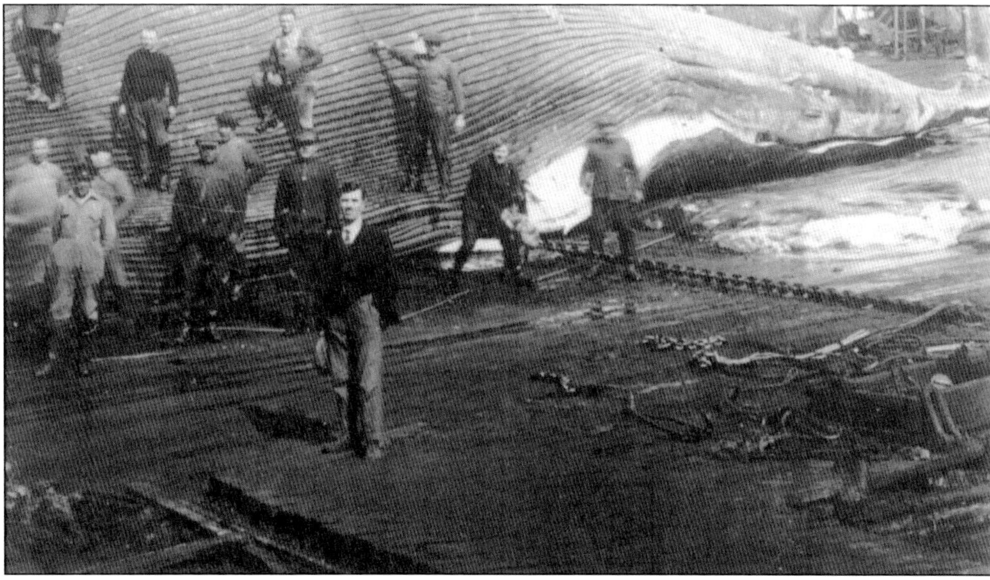

This close-up from the image on the previous page shows the station manager in a coat and tie—he will never be seen on the whale, as he does not have the shoes for it. It takes many people to run a whaling operation, in addition to 11 to 12 people on each of the boats. The factory workers include Norwegians, Eskimos, Americans, Canadians, Germans, Japanese, and others.

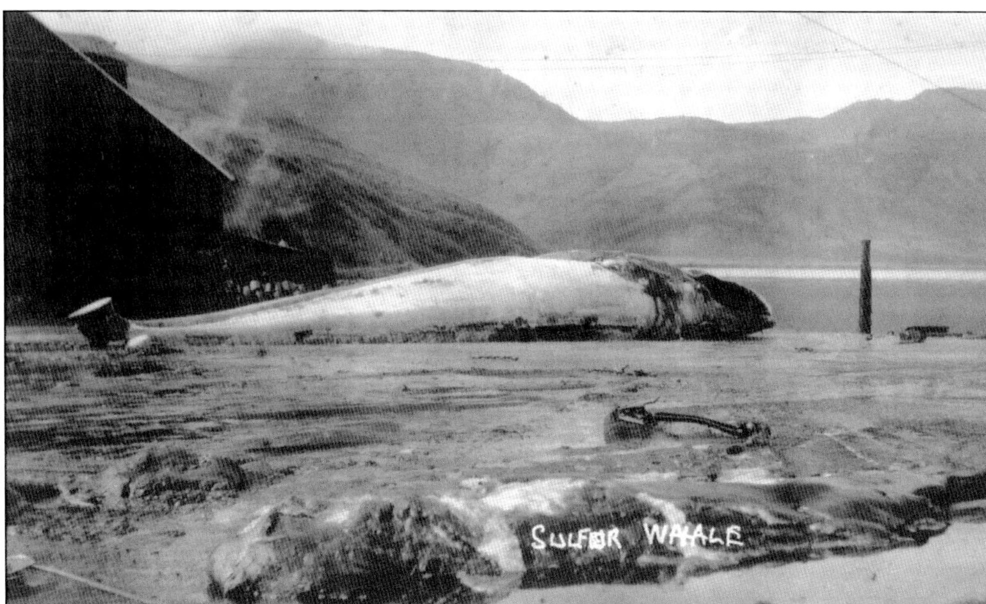

This blue (or "sulfur bottom") whale has already had its outside skin and blubber removed and is ready to be dismembered. The head will be removed, and after the meat is removed, the bones will be separated and broken up so that they fit into the bone grinder to be made into meal.

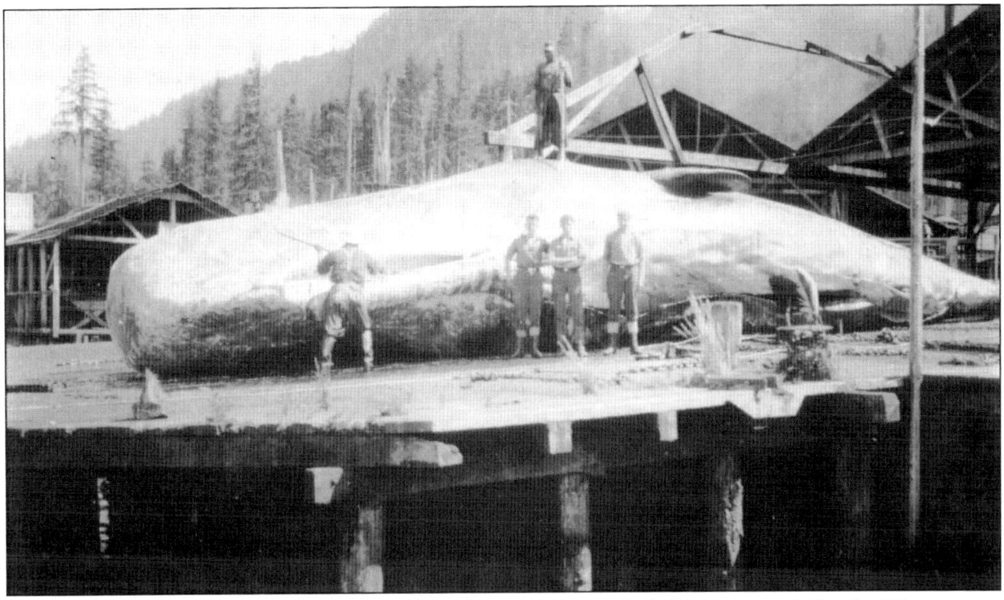

Some men pose while another man works. The is a sperm whale lying on its left side. The station is not identified, it but could be Port Armstrong—nicknamed "Smellstrong" for its pervasive odor. Unfortunately, it was located too close to Juneau and Sitka. The presence of trees in the background usually indicates Port Armstrong, but this could also be the Tyee plant on Baranov Island.

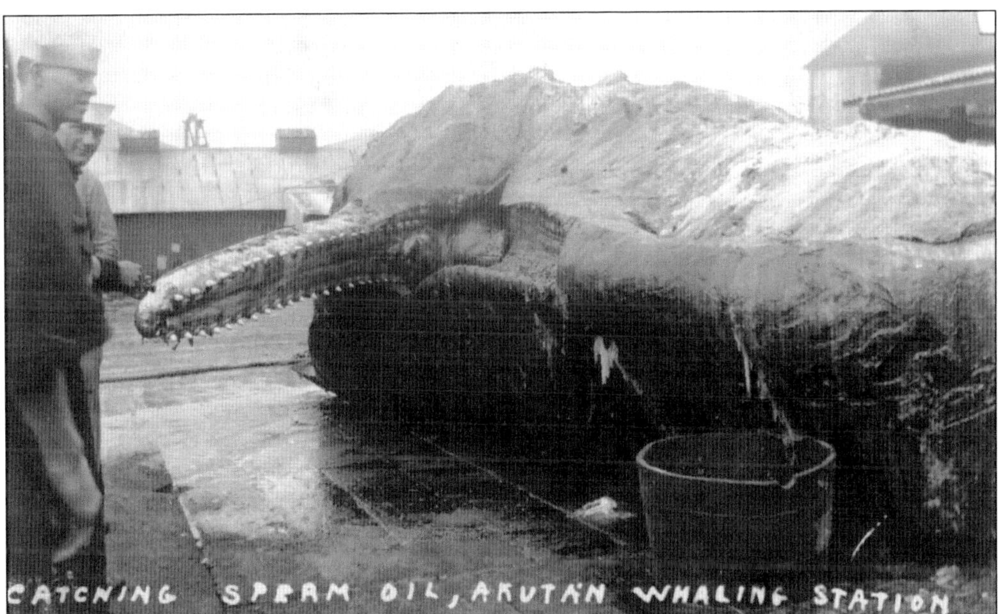

There are many Navy personnel in pictures of Akutan because of the station's close proximity to a naval base. Here, Navy men inspect the business end of a sperm whale; note the two rows of teeth in the jaw.

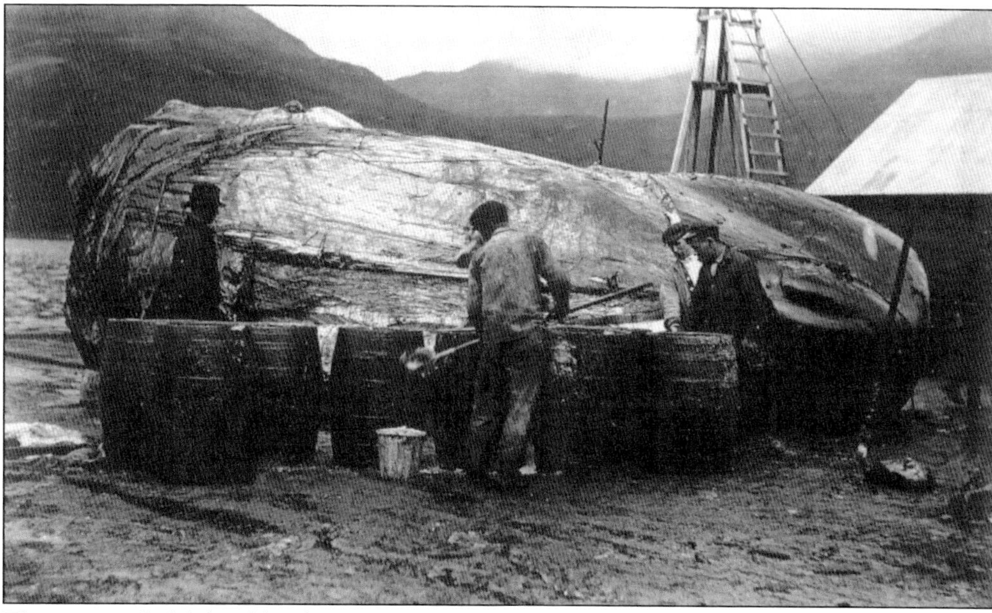

This sperm whale's head has been removed intact from the body of the whale. This is common practice so that the valuable sperm oil can be removed with the least waste. The workers cut a channel in the body, and one is using a scoop to get the oil as it oozes out of the head.

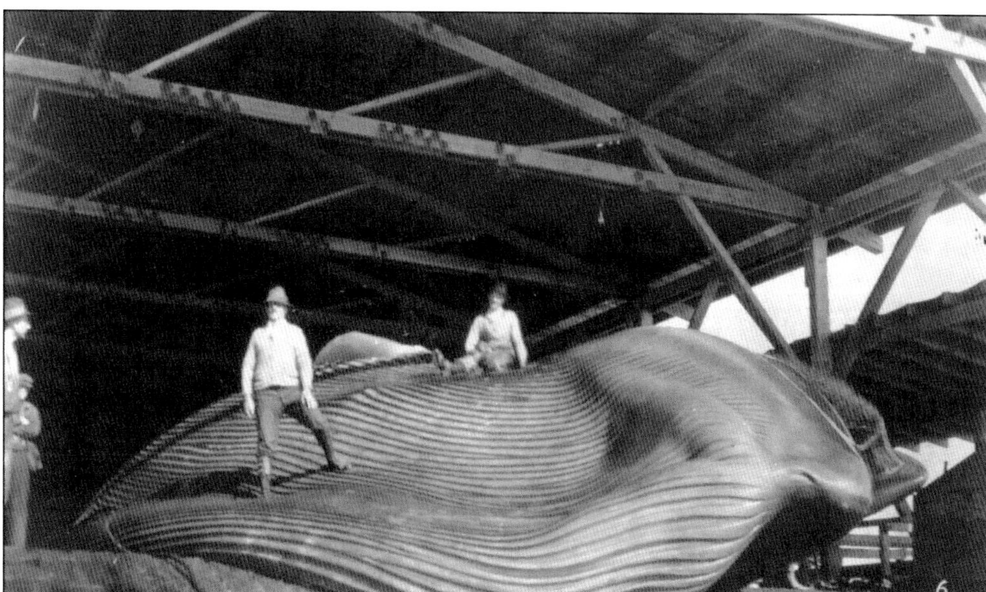

These men pose with a humpback whale on the flensing platform. Note the large fin on top of the whale, with the whale lying on its side; these fins can reach 15 feet in length. This fin length is the characteristic most easily used to identify the humpback whale.

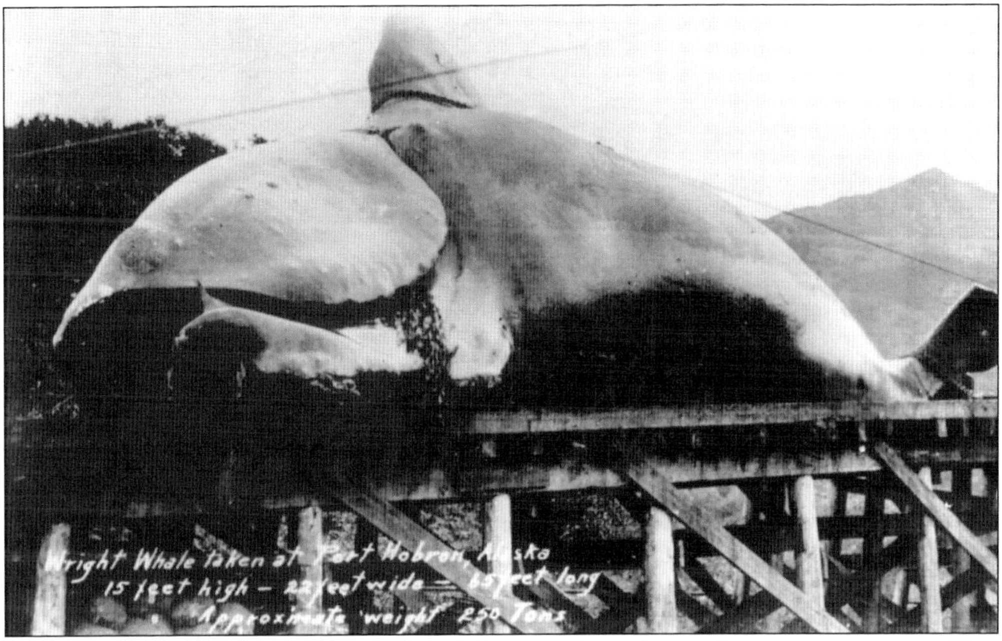

The right whale is very rarely caught by whalers—perhaps they will get one in a season. This one was taken at Port Hobron, Alaska, which is in the Kodiak vicinity. The right whale got its name because it was the "right whale" to harpoon since it would float afterward, while other larger whales would sink after being shot.

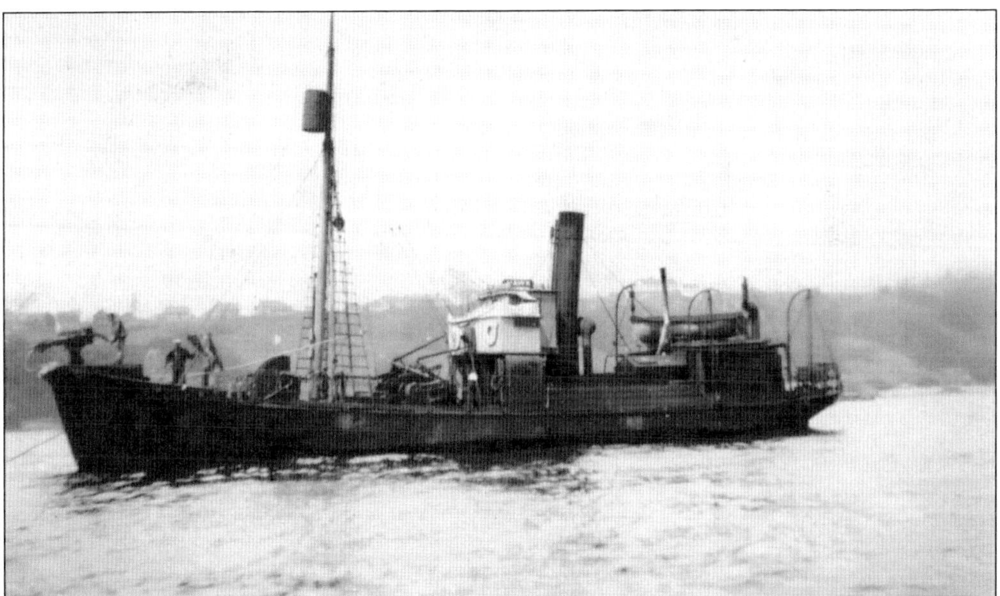

This typical whaling steamer has a barrel on the mast where crew members could keep a lookout for whales; this position offered a greater advantage for spotting the blow of a whale. The ship also contains a manned whale gun on the bow. Most of the whalers were Norwegian, but Alaska was an exception, as many of the boats in the area were American-made in order to comply with certain rules and regulations.

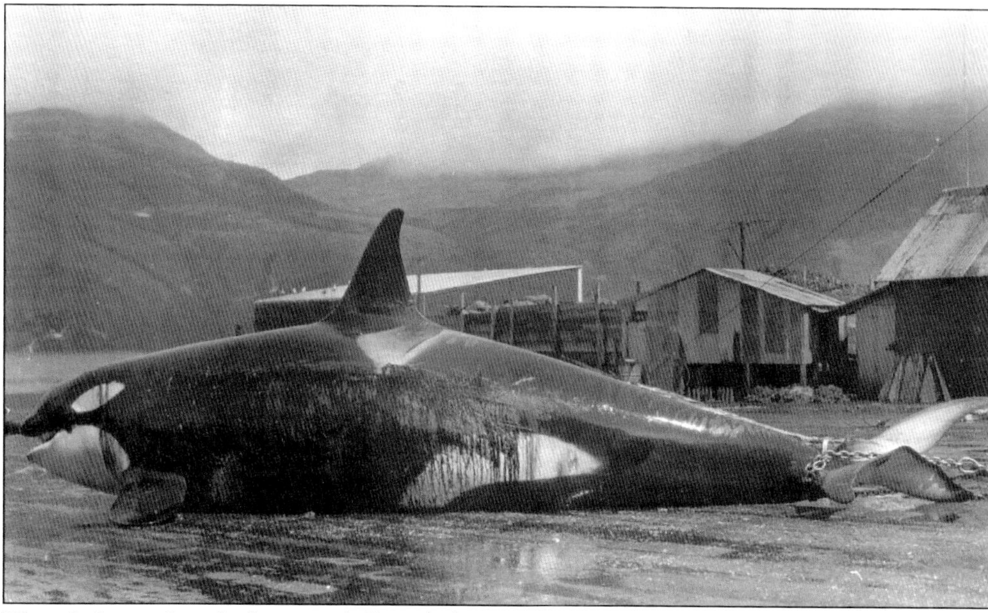

The whale on this platform is a killer whale, or orca, which is seldom taken by whalers. This whale may have been caught at the request of a biologist. Killer whales are top predators known for feeding on seals, but they can also act in unison to killer larger whales.

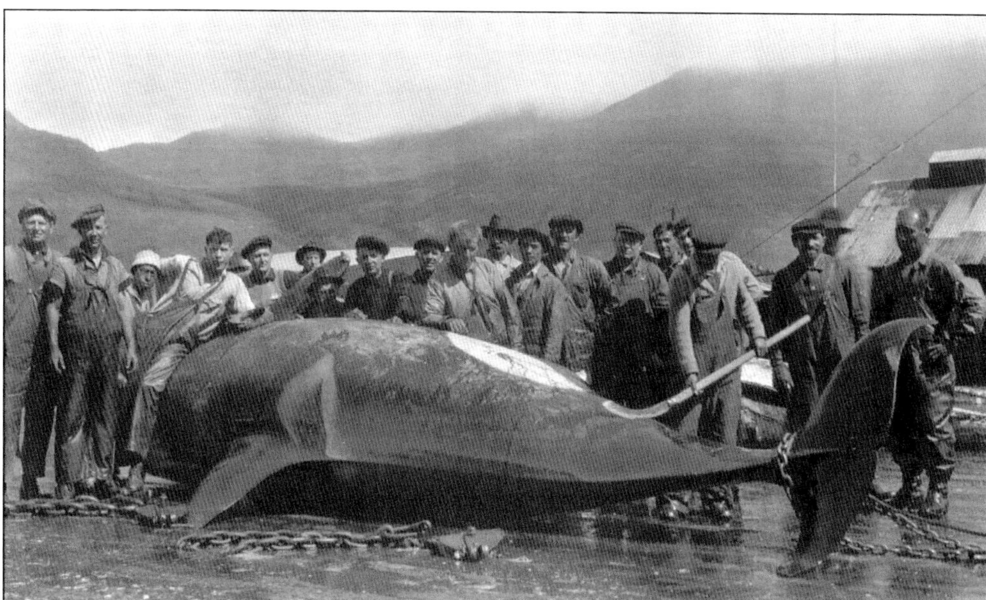

This is the same killer whale from the previous image, now on its left side with station employees in the picture. The head flenser is at right with his large flensing knife. The head flenser is in charge of the flensing platform much like a captain is in charge of his ship.

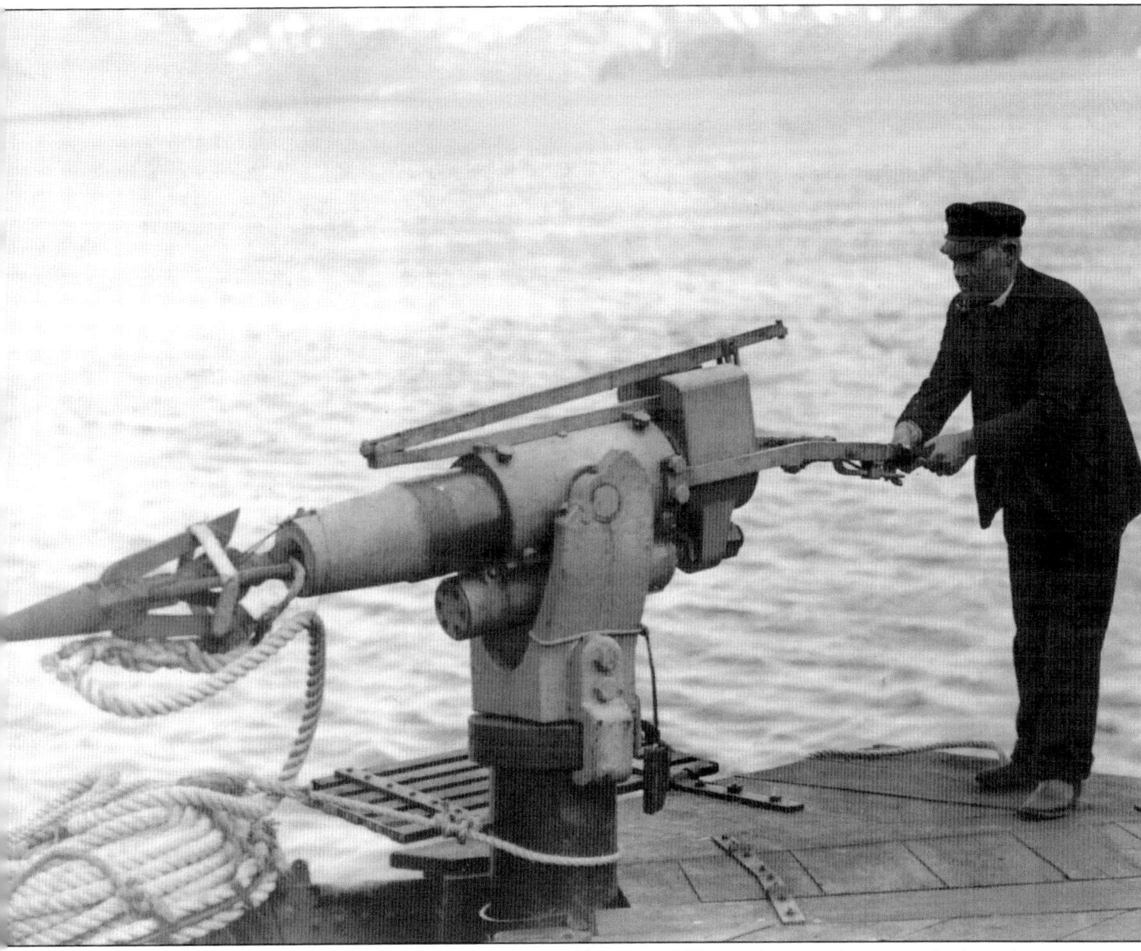

This is a modern whaling gun of the type invented by Svend Foyn. A time fuse attached to the bomb on the harpoon is ignited when the gun is fired, and the bomb explodes five seconds later. Barbs on the shaft of the harpoon expand in the whale, making it next to impossible for the harpoon to draw out.

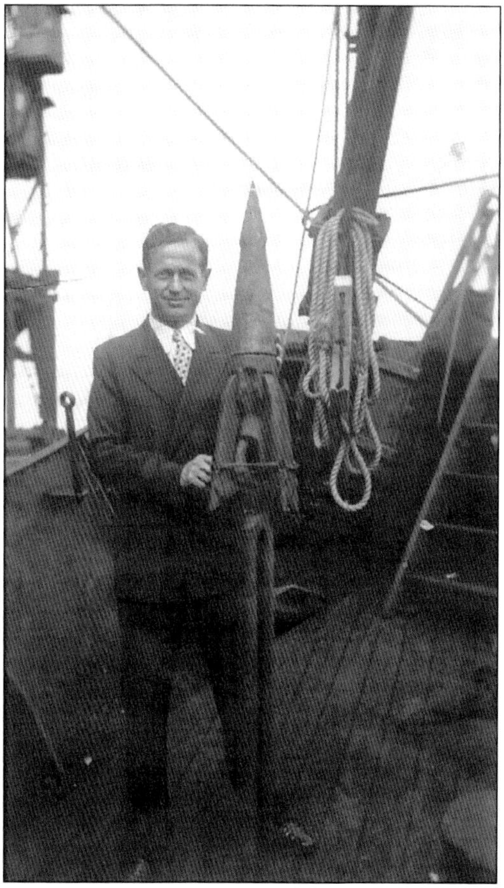

Here is a typical harpoon. This would be put into the cannon on the bow of the whaling steamers. The tip, loaded with dynamite, is the explosive part. Usually, if the shot is good, the whale will die quickly. If the shot is not good and the harpoon does not draw out, the whale can pull the boat around for hours.

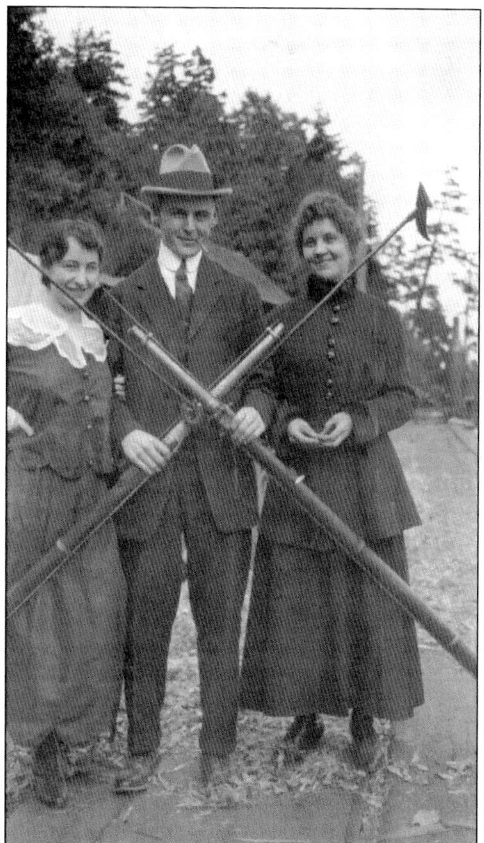

This threesome, attired for the early 1900s, holds a harpoon and darting gun, which is used when in close proximity to the whale. The darting gun also has an explosive head, which has a projectile that is fired when the lance on the end enters the whale. These two and the darting gun in the next picture are still used by the Eskimos today. The larger cannon is not used in modern whaling, except by Japanese hunters in the Antarctic looking for larger whales.

These are the brass guns that fired a projectile about 13 inches in length with an explosive tip. This type of gun is used at close range to kill the whale, and is still used today by Eskimos. Today, the Alaska Eskimo Whaling Commission (AEWC) has purchased the rights and manufactures the guns and darts for the whalers.

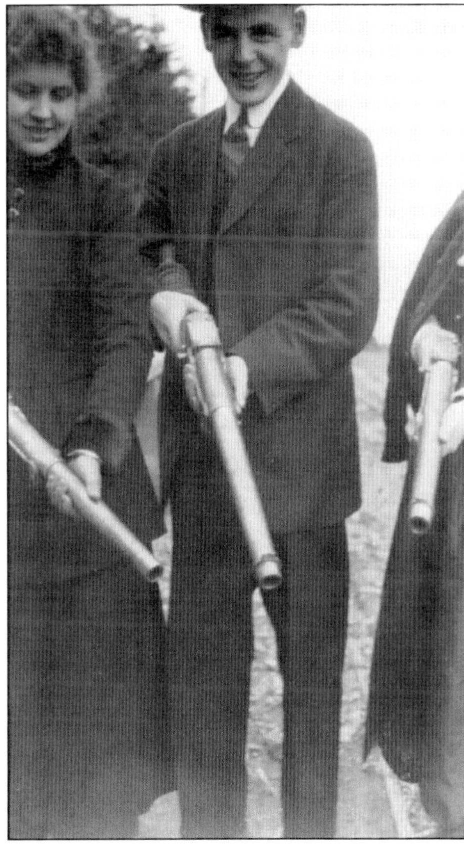

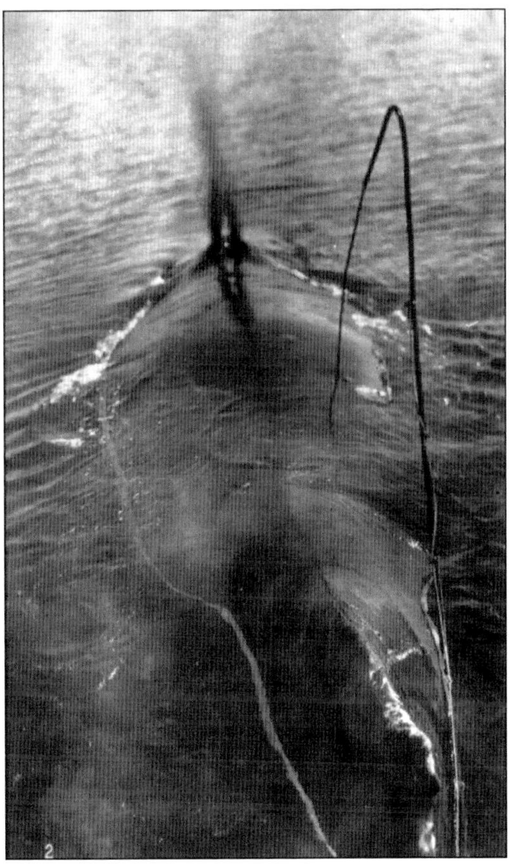

According to the 1909 *Report of the Commissioner of Fisheries*, if the whale blows blood when it reaches the surface, it is mortally wounded, and the whalers wait until it dies, but if it blows clear and is quiet, a pram—a spoon-shaped boat—is rowed alongside the whale and a long lance is driven into it until it blows blood, which shows an internal hemorrhage that indicates that the whale will soon expire, rolling over on its back in its last struggles before it begins to sink.

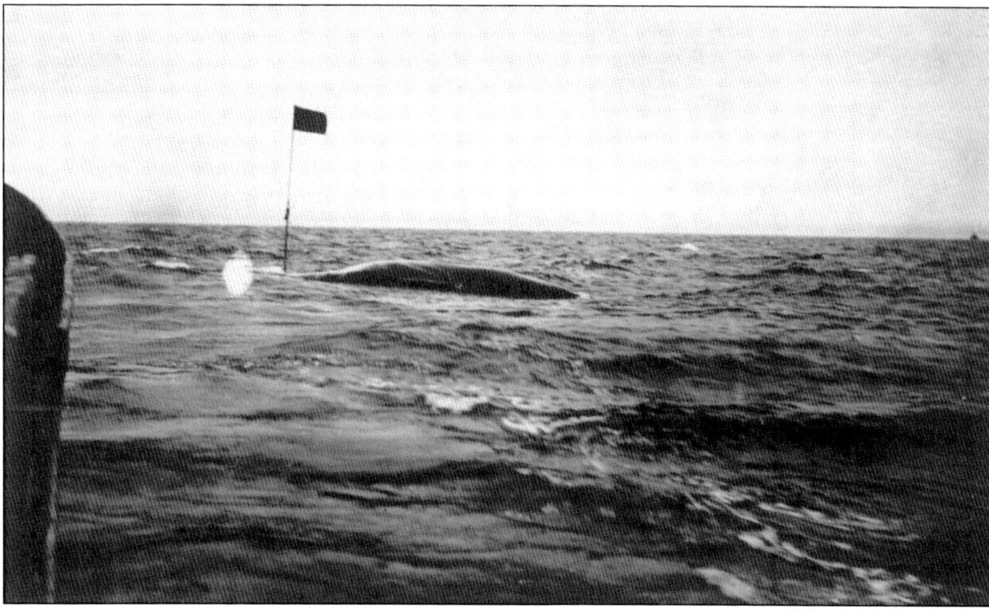

The report continues: "Should the whaler not be ready to return to the station immediately, a buoy with the ship's flag attached is secured to the whale and both are allowed to go adrift while the steamer continues its hunt, sometimes as many as three whales being brought in at one time, all with their tails out of the water and hoisted to the bow."

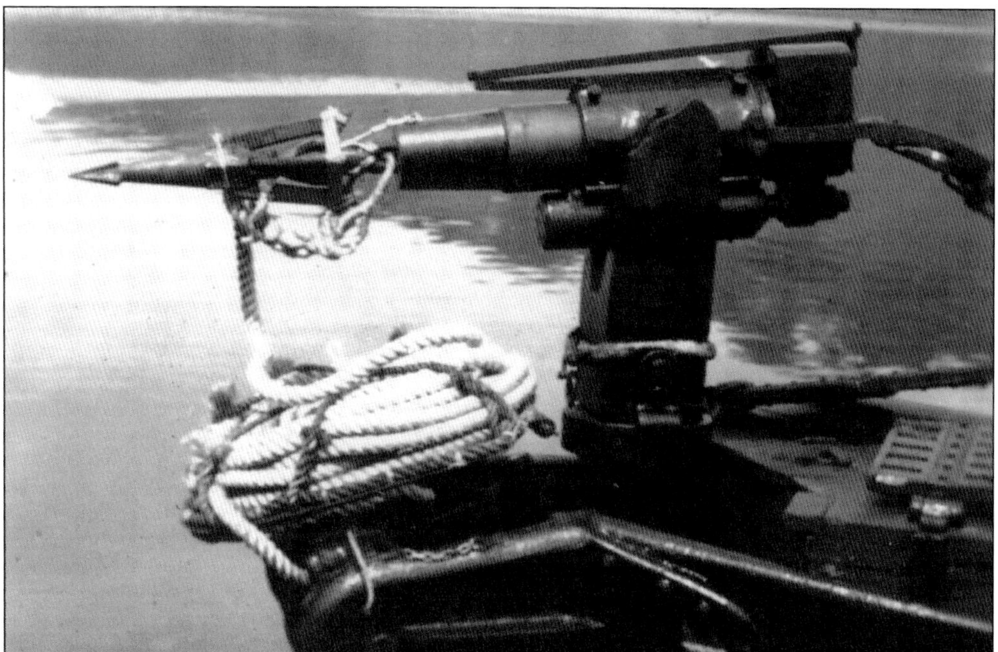

Another type of cannon shows the rope tied to the bow of the ship and the back of the harpoon tip. This will keep the barbs from opening up and cause the harpoon to miss its target, or not go in completely, which could result in a lost whale. The harpoon was designed to keep the barbs from opening up until the harpoon entered the whale. After entering the whale the barbs would open up to prevent the harpoon from withdrawing from the whale.

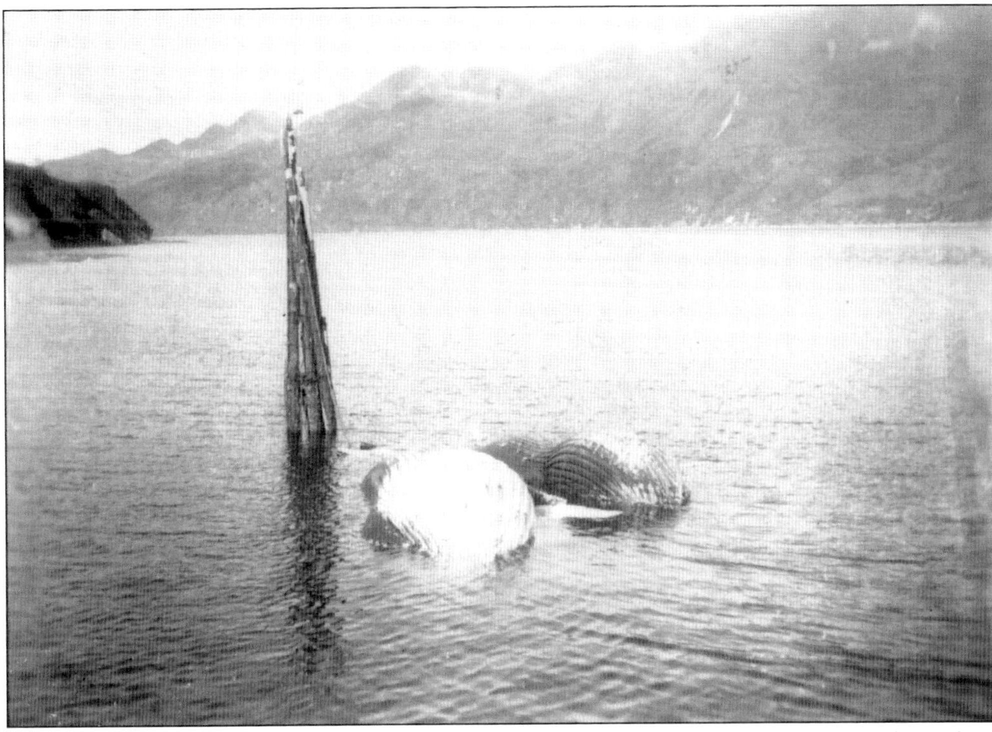

The whaler—or, in some cases, a tug designed for the job—will bring the whales in to be tied out until needed. In either case, the boat and crew will return to whaling as soon as possible.

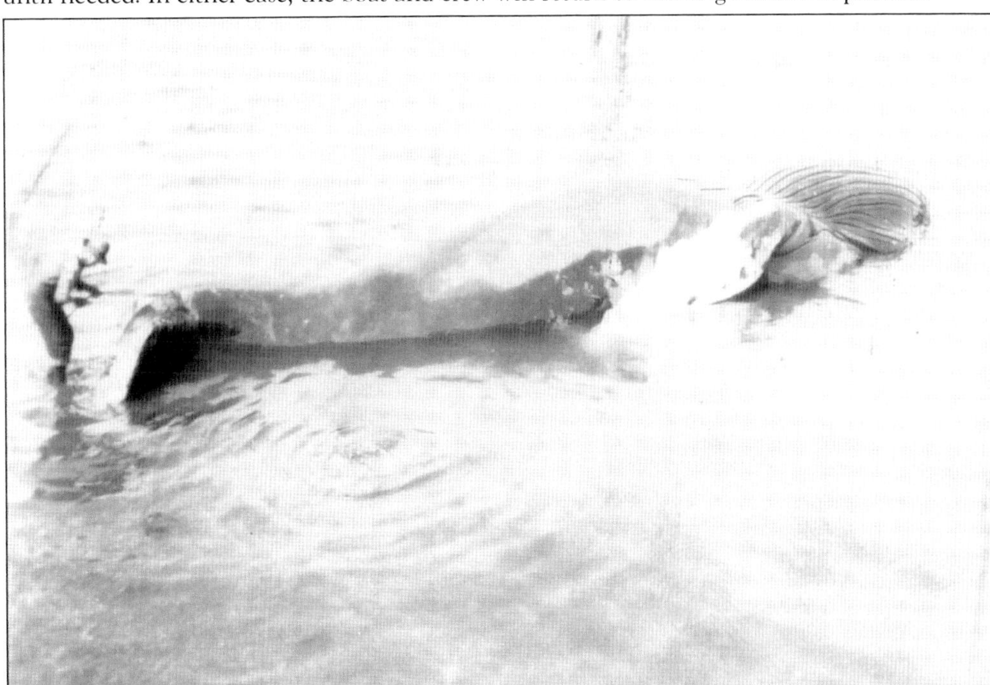

A worker attaches a cable around what is left of the tail of this whale; the tail was cut at the boat so it would not interfere with towing the whale.

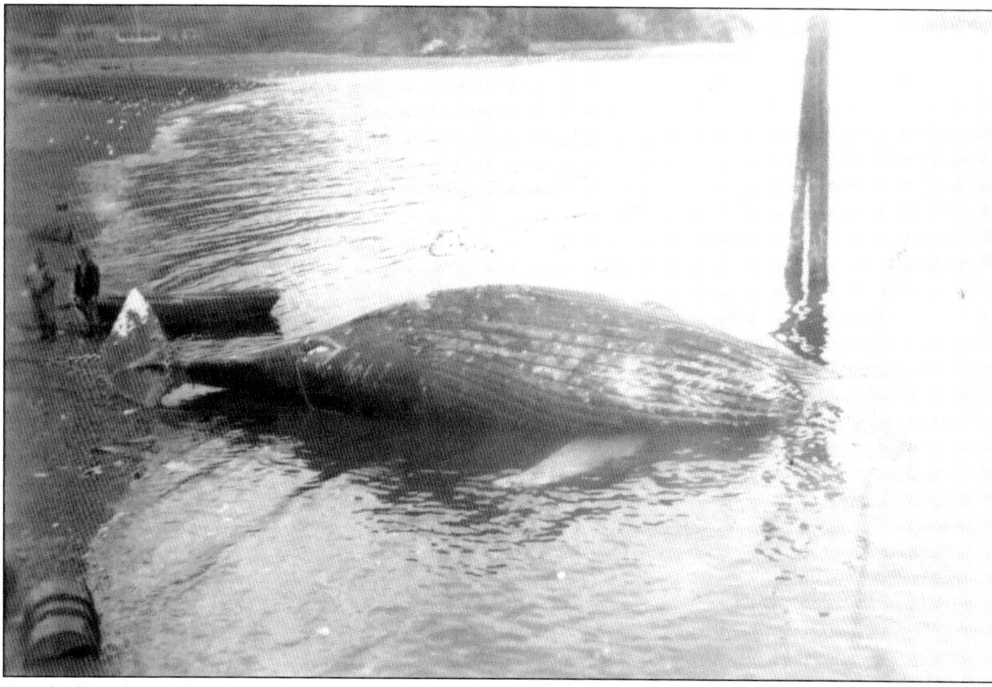

A whale is brought in as far as possible and left partially submerged until the big winches on shore can tow it up the ramp.

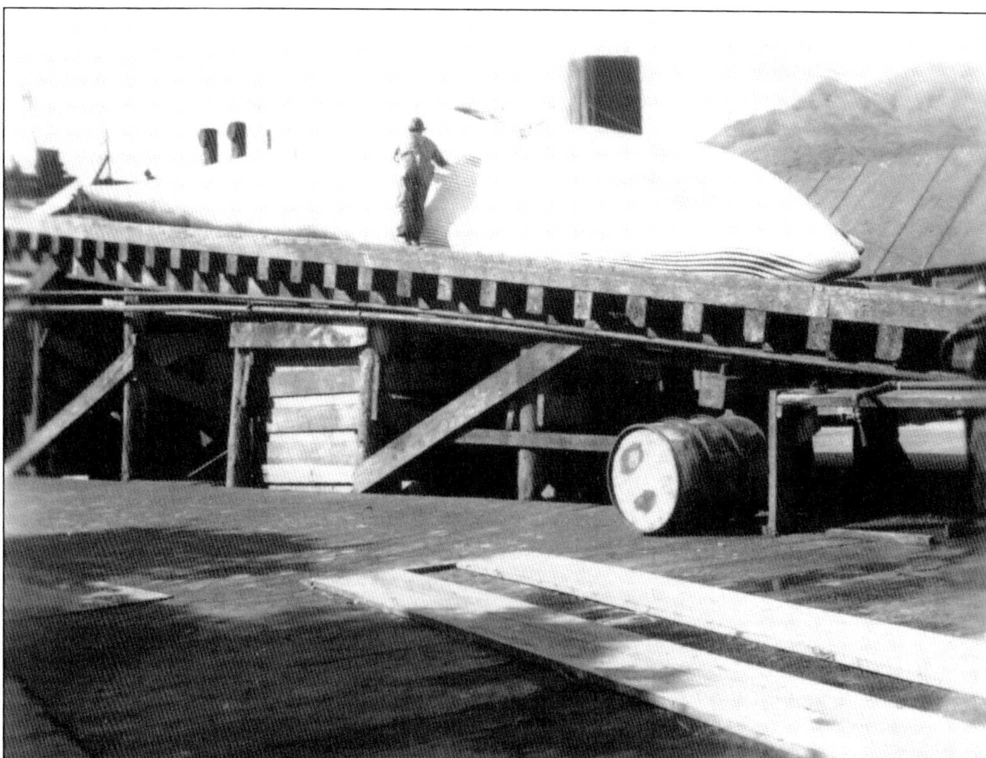

This whale has been hauled up the ramp and is now being prepared for flensing.

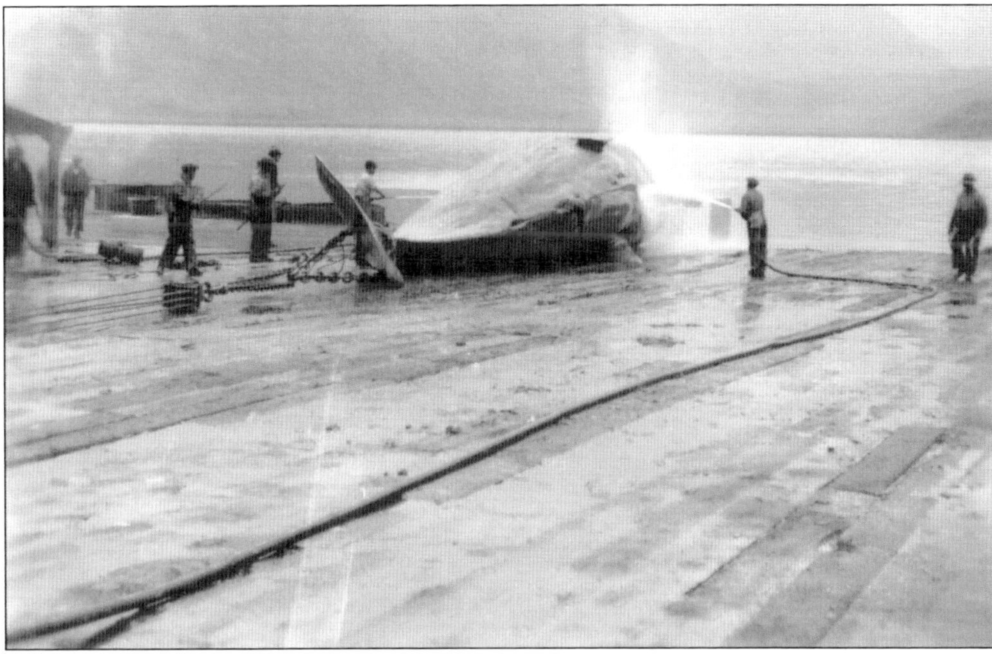

This whale is being given a bath to remove rocks and other debris that may be on it. This will help to keep knives (especially flensing knives) from dulling too early.

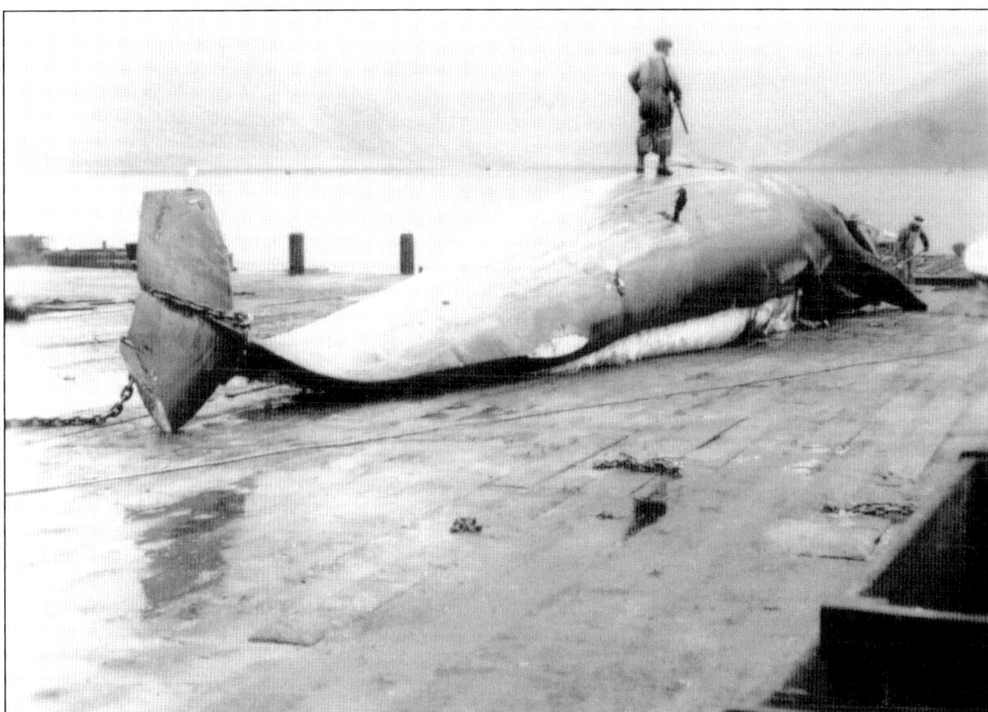

This whale has been cut horizontally, and the flenser, with his long-bladed flensing knife, is preparing to make vertical cuts for removing the blubber.

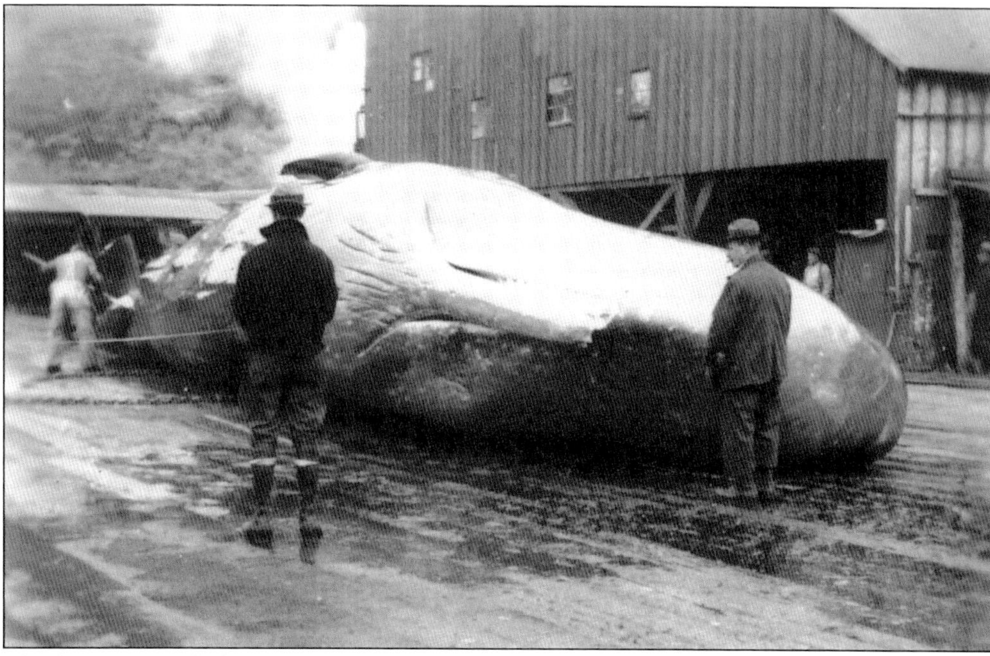

This large sperm whale can be identified by its large bulbous head and its lower jaw with menacing teeth used to catch large squid, which is a major part of its diet. Another identifying mark is the sperm whale's skin—rather than being smooth like the skin of other whales, it appears very wrinkly.

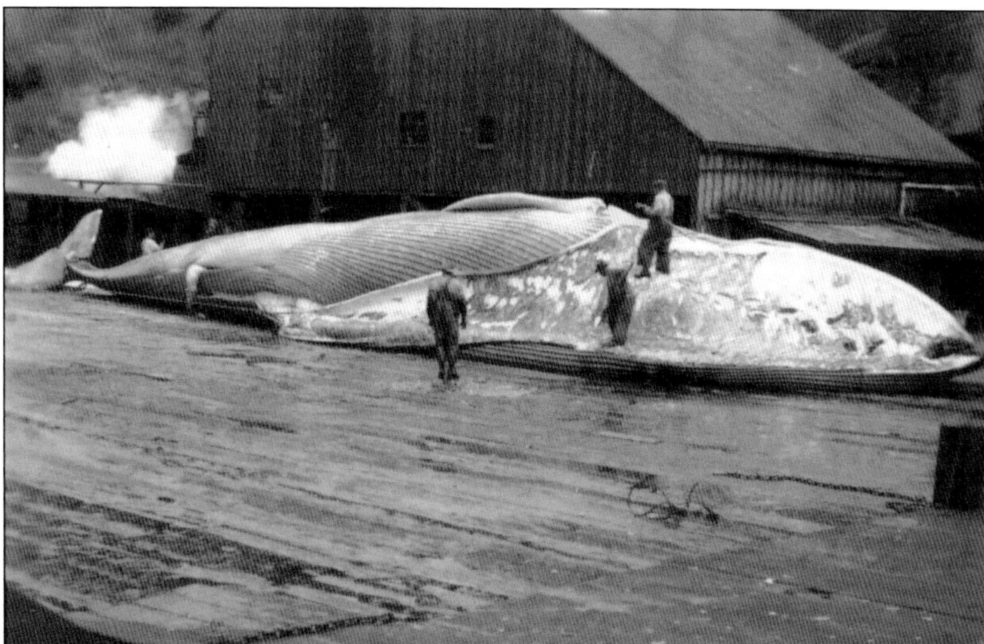

On this carcass, the oily blubber is partially cut away, exposing the meat. According to Cobb and Kutchin: "The layer of fat directly under the skin varies in thickness from 4 to 7 inches on the whales found in southeast Alaska waters."

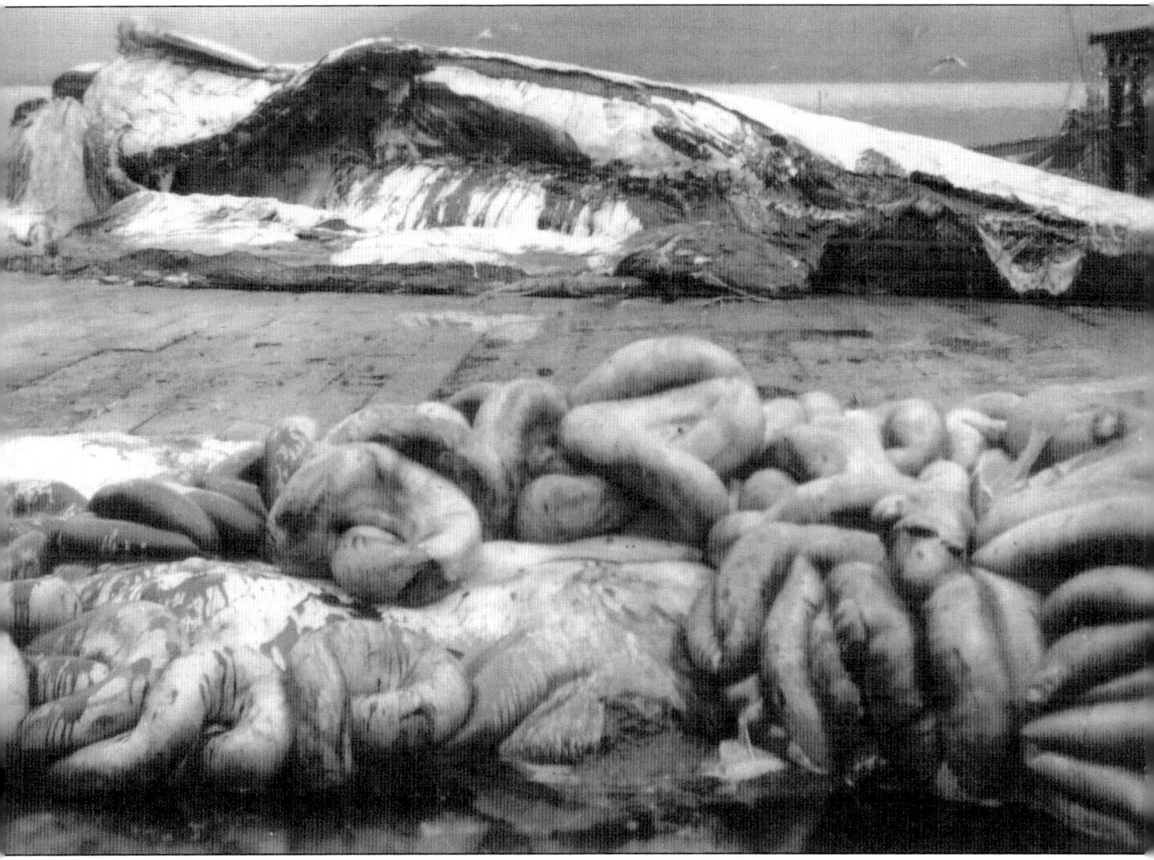

The intestines are drawn out of the body before the carcass is processed. This is important, because if the carcass sits for too much time, the whale will begin to expand from the build-up of gases and may actually explode, spewing stinky meat and fat everywhere.

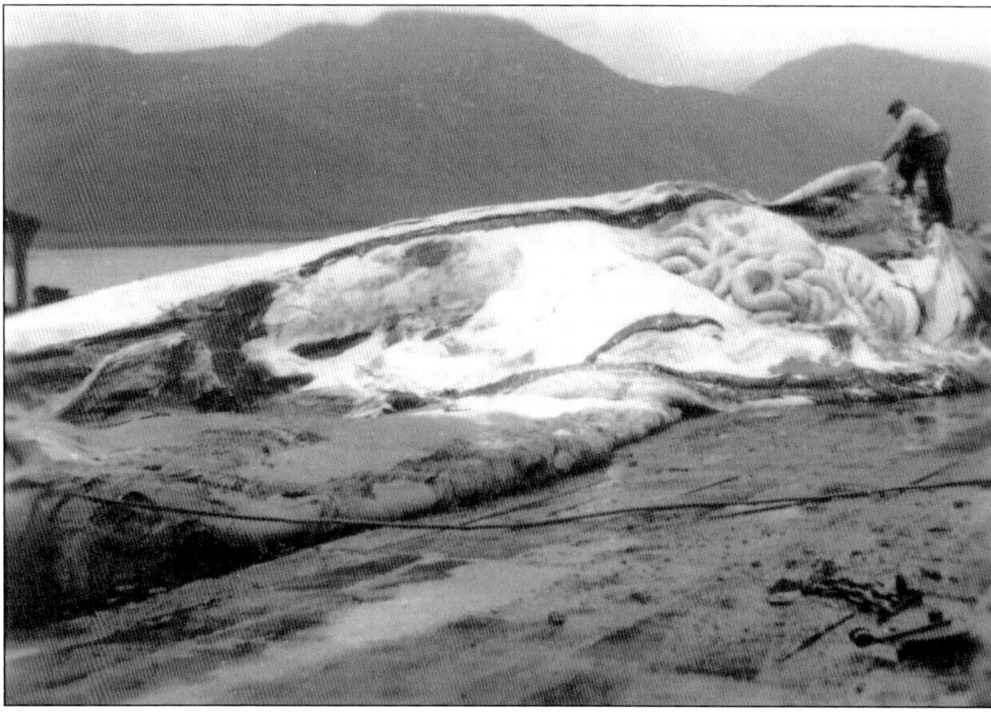

Here, skin and internal organs have been separated from the carcass and will be treated in a separate process.

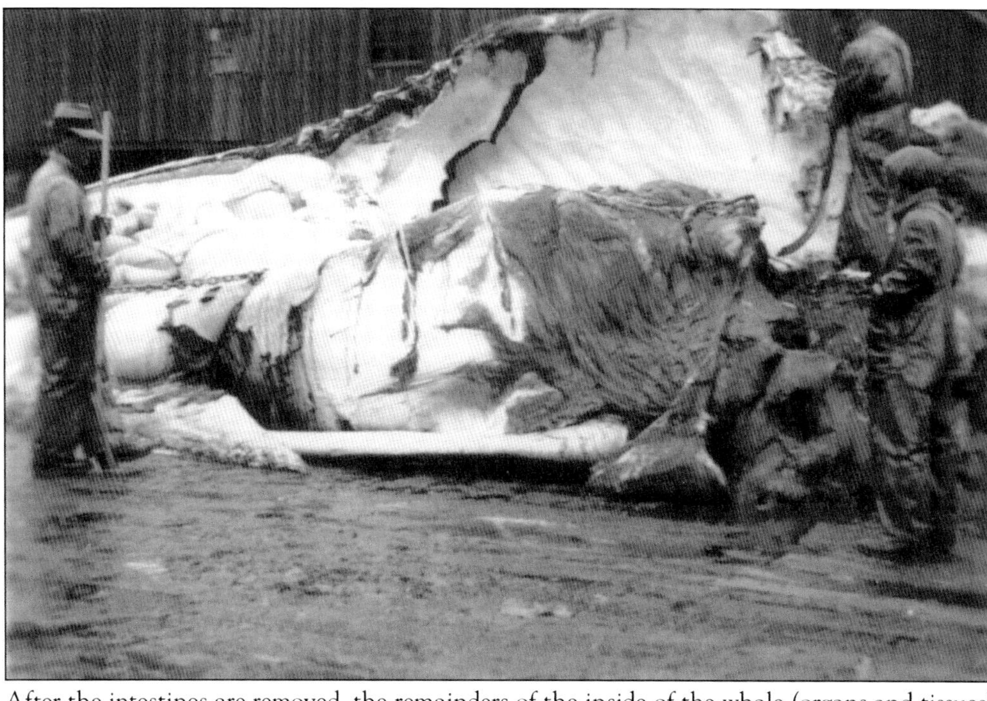

After the intestines are removed, the remainders of the inside of the whale (organs and tissues) are removed and separated from the carcass.

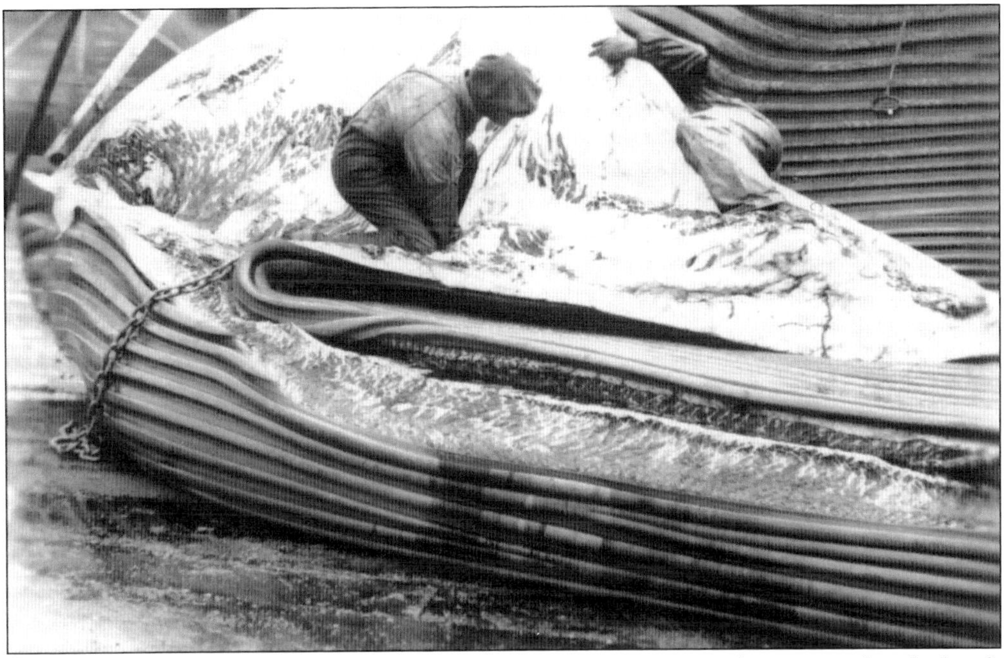

The head of this baleen whale is being cut up for processing. Because there is no oil in the heads of other large whales, there is no special processing except for the sperm whale.

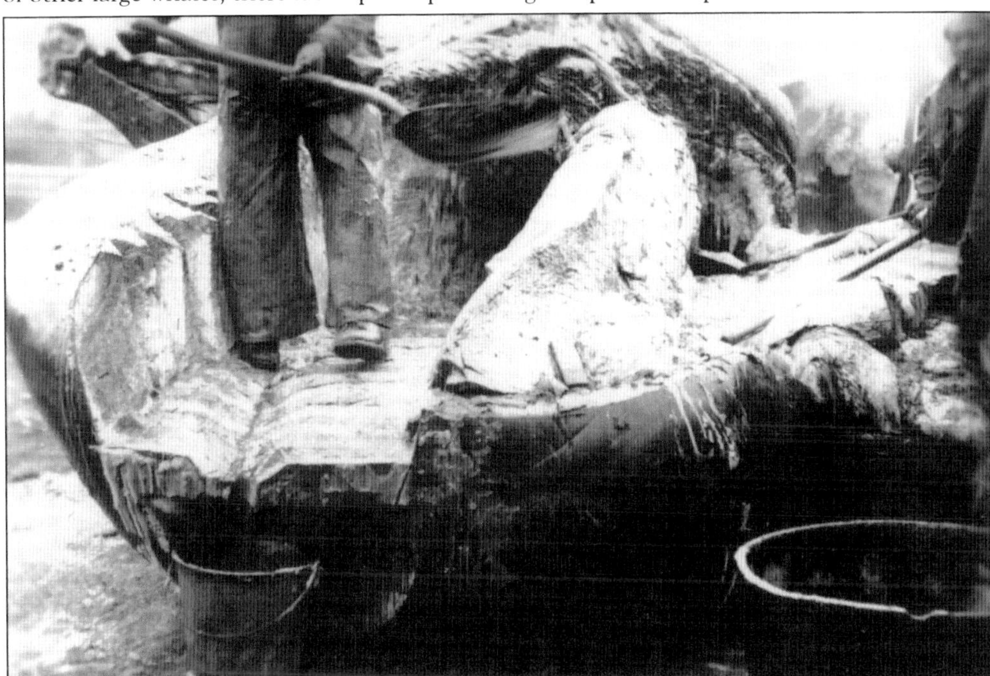

After a sperm whale's head is cut, oil flows freely and is caught "by the bucket" and poured into barrels. The oil made from the head of a sperm whale is the finest machine oil available—or, today, not available. Sperm whale oil and other marine mammal products are no longer available because the United States and other countries passed legislation in 1972 banning their production or importation.

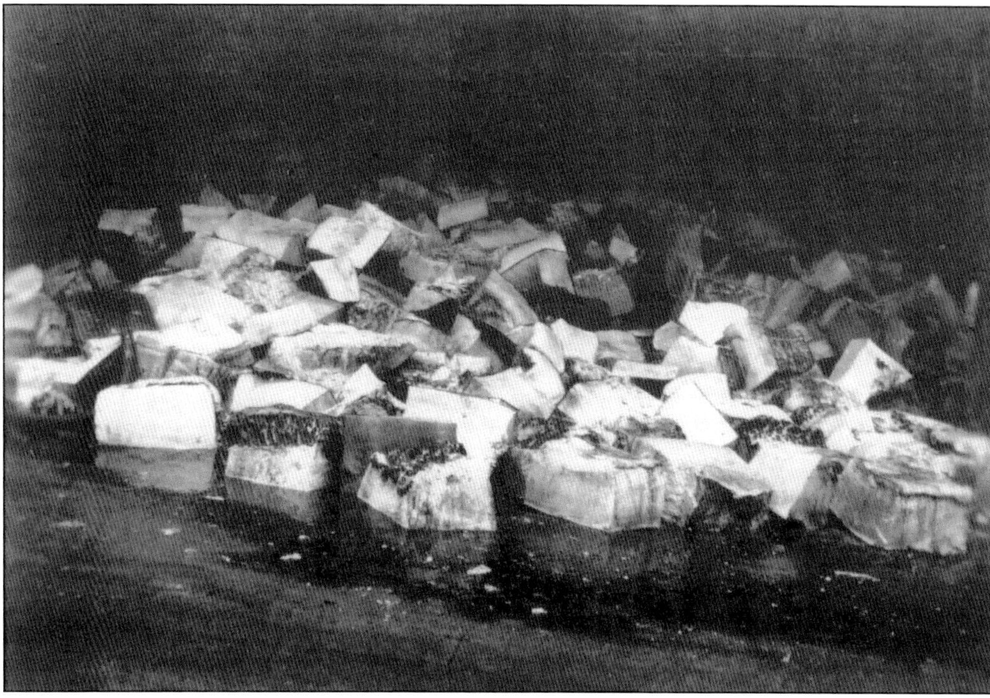

Chunks of whale blubber are ready for the rendering vats. In the past, these were sliced on one side while still being held together like a book. These were actually call bible-leaves.

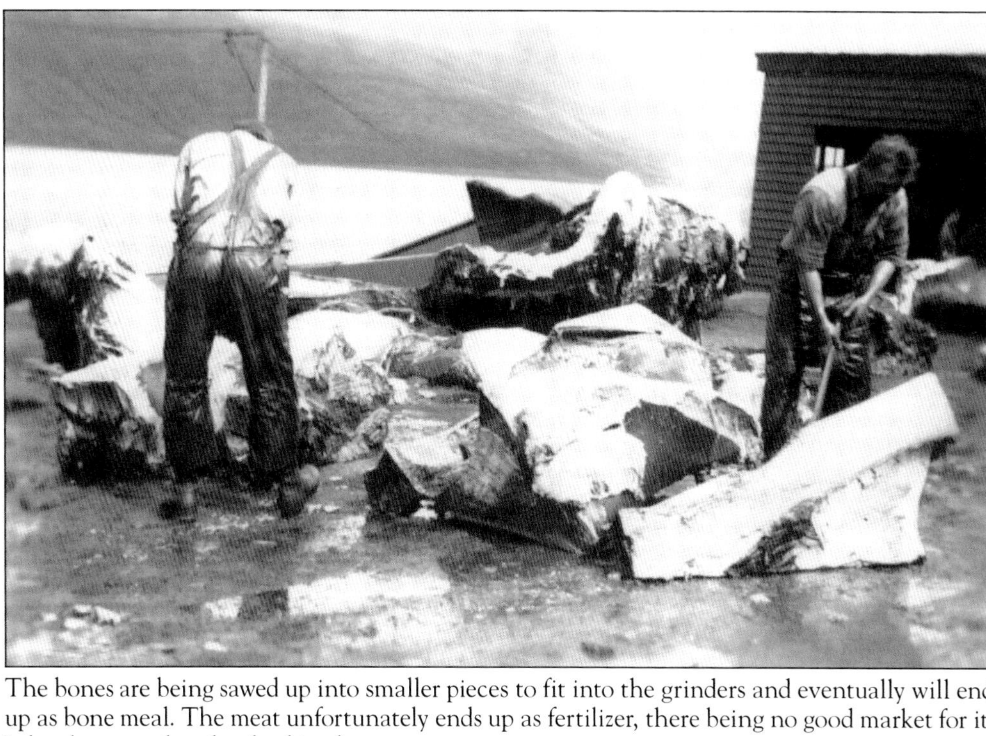

The bones are being sawed up into smaller pieces to fit into the grinders and eventually will end up as bone meal. The meat unfortunately ends up as fertilizer, there being no good market for it. It has been used as dog food in the past.

BIBLIOGRAPHY

Cobb, John Nathan, and Howard Malcolm Kutchin. *Fisheries of Alaska*. Washington, DC: US Government Printing Office, 1911.

Report of the Commissioner of Fisheries. Washington, DC: Department of Commerce and Labor, Bureau of Fisheries, 1909, 1916.

Tønnessen, J.N., and A.O. Johnsen. *The History of Modern Whaling*. Translated from the Norwegian by R.I. Christophersen. Berkeley and Los Angeles: UC Press, 1982

Webb, Robert Lloyd. *On the Northwest: Commercial Whaling in the Pacific Northwest 1790–1967*. Vancouver: University of British Columbia Press, 1988.

VanStone, James W. Point Hope: *An Eskimo Village in Transition*. Seattle: UW Press, 1962.

Vinnedge, Dale A. *Pacific Fisherman Magazine Whaling Articles: A Compendium of 975 Whaling Articles from Pacific Fisherman Magazine, 1905–1972*. Unpublished, 1994–2002.

Consistent with our mission to preserve history on a local level, this book was printed in South Carolina on American-made paper and manufactured entirely in the United States. Products carrying the accredited Forest Stewardship Council (FSC) label are printed on 100 percent FSC-certified paper.